2nd International Conference on Information Processing and Control Engineering (ICIPCE 2018)

MATEC Web of Conferences Volume 214 (2018)

Shanghai, China
27 – 29 July 2018

Editor:

G. Beydoun

ISBN: 978-1-5108-7312-4

Printed from e-media with permission by:

Curran Associates, Inc.
57 Morehouse Lane
Red Hook, NY 12571

proceedings.com

Some format issues inherent in the e-media version may also appear in this print version.

Printed by Curran Associates, Inc. (2018)

For additional information, please contact EDP Sciences – Web of Conferences at the address below.

EDP Sciences – Web of Conferences
17, Avenue du Hoggar
Parc d'Activité de Courtabœuf
BP 112
F-91944 Les Ulis Cedex A
France

Phone: +33 (0) 1 69 18 75 75
Fax: +33 (0) 1 69 28 84 91

contact-edps@webofconferences.org

Additional copies of this publication are available from:

Curran Associates, Inc.
57 Morehouse Lane
Red Hook, NY 12571 USA
Phone: 845-758-0400
Fax: 845-758-2633
Email: curran@proceedings.com
Web: www.proceedings.com

TABLE OF CONTENTS

Preface

This volume contains papers presented at the 2nd International Conference on Information Processing and Control Engineering, which was held during July 27-29, 2018 in Shanghai, China.

ICIPCE provides a scientific platform for both local and international scientists, engineers and technologists who work in all aspects of Information Processing and Control Engineering. In addition to the contributed papers, internationally known experts from several countries are also invited to deliver keynote and plenary speeches at ICIPCE 2018.

The volume includes 23 selected papers which were submitted to the conference from universities, research institutes and industries. Each contributed paper has been peer-reviewed by reviewers who were collected organizing and technical committee members as well as other experts in the field from different countries. The proceedings tend to present to the readers the newest researches results and findings in the field of Information Processing and Control Engineering.

Much of the credit of the success of the conference is due to topic coordinators who have devoted their expertise and experience in promoting and in general co-ordination of the activities for the organization and operation of the conference. The coordinators of various session topics have devoted a considerable time and energy in soliciting papers from relevant researchers for presentation at the conference.

The chairpersons of the different sessions played important role in conducting the proceedings of the session in a timely and efficient manner and the on behalf of the conference committee, we express sincere appreciation for their involvement. The reviewers of the manuscripts, those by tradition would remain anonymous, have also been very helpful in efficiently reviewing the manuscripts, providing valuable comments well within the time allotted to them. We express our sincere and grateful thanks to all reviewers.

ICIPCE 2018 Organizing Committee

July 27-29, 2018

Statement of Peer review

In submitting conference proceedings to MATEC Web of Conferences, I certify to the Publisher that I adhere to the **Policy on Publishing Integrity** of the journal in order to safeguard good scientific practice in publishing.

1. All articles have been subjected to peer review administered by the proceedings editors.
2. Reviews have been conducted by expert referees, who have been requested to provide unbiased and constructive comments aimed, whenever possible, at improving the work.
3. Proceedings editors have taken all reasonable steps to ensure the quality of the materials they publish and their decision to accept or reject a paper for publication has been based only on the merits of the work and the relevance to the journal.

Title, date and place of the conference

2018 2nd International Conference on Information Processing and Control Engineering

Shanghai, China, July 27-29, 2018

Proceedings editor(s):

Guest Editor

Prof. Ghassan Beydoun, University of Technology Sydney, Australia

Ghassan Beydoun

Date and editor's signature

28/09/2018

Design Scheme of Communication Node Based on FlexRay Bus

Xin He[1], Jia'nan Wu[1], Hongde Deng[1], Zean Zhen[1] and Chenyang Liu[1]

[1]*School of Automation, Northwestern Polytechnical University, China*

Abstract. As a high-speed and reliable serial communication bus, the FlexRay bus has gradually become the mainstream of vehicle network systems. In order to verify the feasibility of the bus technology in aerospace applications, this paper designs a FlexRay communication node based on TI's high-performance microcontroller TMS570LS3137 for security systems. In the whole system, with the DSP of core, the FlexRay communication protocol is achieved by using the bus controller TJA1080 to carry out data transmission. This method is suitable for data transmission of pint and medium-sized UAV (unmanned air vehicles) flight control system with high integration and high transmission rate.

1 Introduction

Aviation bus [1-2] is one of the basic and key technologies of avionics system. With the development of airborne avionics technology, higher requirements are proposed for the airborne data bus. At present, the data bus is widely used on modern flight control computers to realize data exchange and sharing among its various subsystems. While improving the reliability, expansibility and flexibility of the airborne electronic system, the bus performance indirectly determines the flight performance and flight safety of UAV (unmanned air vehicles). Therefore, it is an important means for the research on bus technology to improve the performance of the flight control system.

Currently, among the communication bus widely used in the aviation field, the communication rates of the 1553B, ARINC659, and other buses all meet the requirements, but the cost is too high. The transmission rate of the ARINC429 bus is only 100 Kbps [3-4], which is not suitable to use in pint and medium-sized UAV. The CAN bus has been widely used due to its reliable performance and strong anti-interference ability, and the real-time and reliability of data communication have also been fully verified [5]. However, with the rapid development of new type UAV, the CAN bus cannot meet the index requirements of increasingly complex system, due to its limitations of communication rate and communication mechanism. Thus, it is necessary to find a new bus technology with high data transmission rate and high security to meet the large-capacity and real-time communication requirements of new type UAV flight control system.

The new generation of vehicle bus FlexRay [6-7] has high transmission rate, high reliability, high security and low cost, which data transmission rate up to 10Mbps, efficiency up to 96.9%. As its performance is not lower than other mainstream aviation bus, it is an inevitable trend to adopt it as the communication network of new type UAV. This paper designs a communication node for pint and medium-sized UAV flight control system, which based on the technology of FlexRay bus.

2 Design of FlexRay bus node

2.1. Node architecture

There are three kinds of architecture modes for FlexRay node: MCU (microprocessor), CC (communication controller) and BD (bus driver), MCU (microprocessor) with CC (communication controller) built-in and BD (bus driver), MCU (microprocessor) with CC (communication controller) and BD (bus driver) built-in. The current commonly used are MCU + CC + BD and MCU + BD [8-9]. This system adopts MCU + BD structure, as shown in Fig. 1, among them, the MCU adopts TI Company's TMS570LS3137, and the bus driver adopts NXP Company's TJA1080.

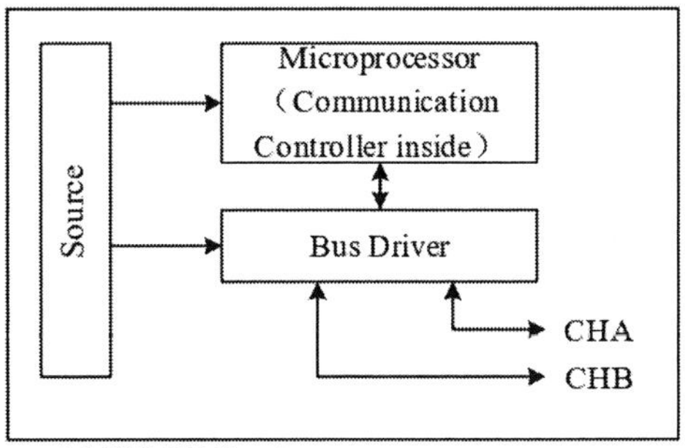

Figure 1. MCU+BD Structure.

2.2 Microprocessor selection

According to the demand of the bus node, it has great influence on security in system. The Hercules series in TI (Texas Instrument), can meet this requirements, which include three kinds of MCU (RM4x, TMS570 and TMS470M) based on ARM Cortex. Among them, the TMS570 is the most suitable for the aerospace fields, railway signal process, etc. At last, TMS570LS3137PGE is selected in this system.

2.3 Design of the whole system structure

The MCU sets its working status by configuring the relevant registers in the communication controller, and realizes the FlexRay bus data transmission by setting the FlexRay protocol status. The bus driver connects the communication controller and the FlexRay bus, and the relevant data in the communication controller is transferred to the bus through the encoding mechanism, and the received data on the bus transmitted to the communication controller through the decoding mechanism. Then, the FlexRay bus communication is realized.

As a node on the FlexRay bus, this system sends signals to the bus and receives the signals transmitted on the bus, and then the related communication tasks are completed. The main task of the FlexRay bus node is data transmission. The whole system consists of TMS570LS3137, TJA1080, SCI and JTAG, etc, as shown in Fig. 2.

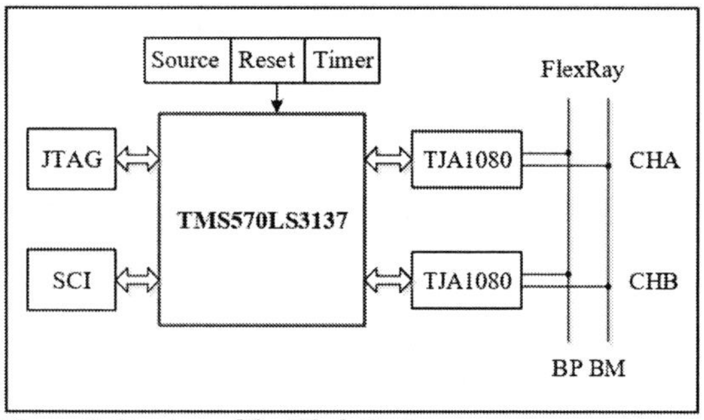

Figure 2. Structure of the Whole System.

3 Key devices

3.1. TJA1080

TJA1080 is a FlexRay bus driver compliant with the FlexRay protocol version 2.1. It is designed to meet the high-speed communication system from 1Mb to 10Mb, and is optimized for time-triggered communication protocols, which is to provide the physical interfaces between modules and channels, and provide differential transmit and receive. It is very flexible and can be used as a node transceiver or an active star transceiver, with excellent ESD (electro-static discharge) protection and power management capabilities [10]. In node mode, the TJA1080 transceiver operates as a single device. In star mode, the transceiver acts as a branch of the FlexRay active star network.

In the power-up stage, if the TRxD0 and TRxD1 are grounded, the transceiver can be configured in a node mode and the following operating modes are selectable:

- Normal (normal power mode)
- Receive-only (normal power mode)
- Standby (low power mode)
- Go-to-sleep (low power mode)
- Sleep (low power mode)

In the power-up stage, if the TRxD0 and TRxD1 are connected to the VBUF, the transceiver can be configured in a star mode and the following operating modes are available:

- Star-idle (normal power mode)
- Star-transmit (normal power mode)
- Star-receive (normal power mode)
- Star-sleep (low power mode)
- Star-standby (low power mode)
- Star-locked (normal power mode)

This paper set it to node mode. Under node configuration, TJA1080 provides two control pins EN and STBN to select the operating mode, its pin function is as shown in the Fig. 3.

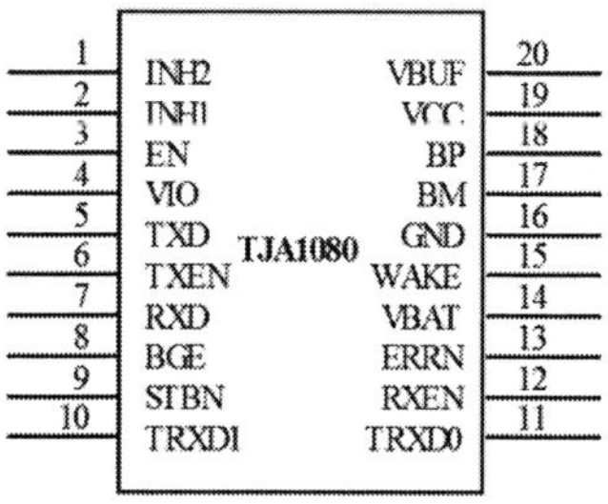

Figure 3. Pin Configuration.

3.2 ACT1210

The essence of the FlexRay signal is a differential signal, so the common mode signal on the bus is the interference signal. Therefore, it is necessary to add a common mode choke in the design. The two coils of the common mode choke are wound on the same core with the same number of turns and the reversed twining direction. In this way, as the differential mode current passes through the common mode choke, the current generates reverse magnetic field in the coil and cancel each other out. And when the common mode current passes through the common mode choke, a co-directional magnetic field is generated in the coil to increase the inductance of the coil, then the coil for common mode signal is characterized by high impedance state, and hinder the common mode signal.

Actually, it is equivalent to a bi-directional filter. On the one hand, it eliminates the common mode electromagnetic interference transmitted over the bus. On the other hand, it controls its own circuit not radiate electromagnetic waves to the outside, to avoid affecting other electronic devices. Common mode choke can transmit differential mode signals, even the signals of direct current and very low frequency can pass. However, it exhibits high impedance for high-frequency common mode noise, and it can be used to inhibit the common mode interference, reduce the electromagnetic radiation, and enhance the EMC (electromagnetic compatibility) of the circuit [10].

The common mode choke in this paper is TDK Company's ACT1210, which is common to the FlexRay bus and CAN bus. It is the same as the winding common mode choke, but realizing the miniaturization. The appearance and its internal circuit are shown in the Fig. 4.

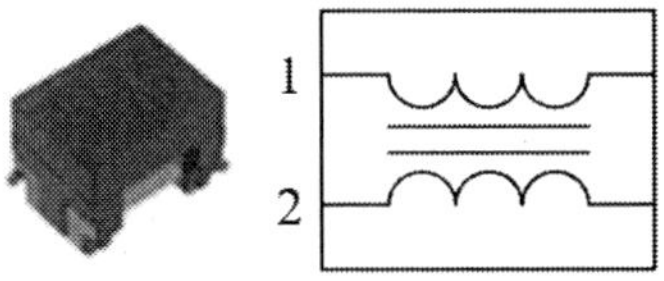

Figure 4. ACT1210.

4 Driving circuit design

The connection between TMS570LS3137 and TJA1080 is the main component of the system hardware design, which is the bridge between the microprocessor and the FlexRay bus. The TMS570LS3137 integrates dual-channel FlexRay bus communication controllers and 8KB message RAM, and the FlexRay controller uses a dual-channel serial, fixed time base multi-master communication protocol with communication rate of 10Mbps per channel.

The driving circuit is a bridge connecting the MCU and the FlexRay bus node, which transmits instruction to the FlexRay bus node. Moreover, the filtering isolation circuit is added before the bus node, which increases the stability and anti-interference ability during the data transmission. The circuit schematic diagram is shown in the Fig.5.

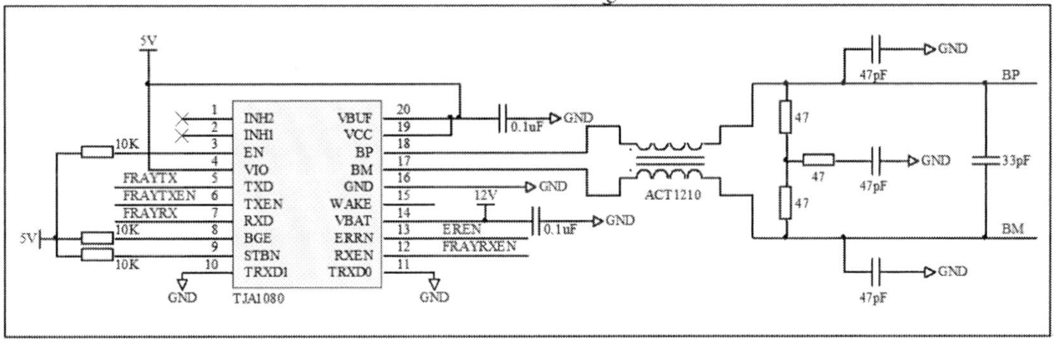

Figure 5. Driving Circuit Schematic.

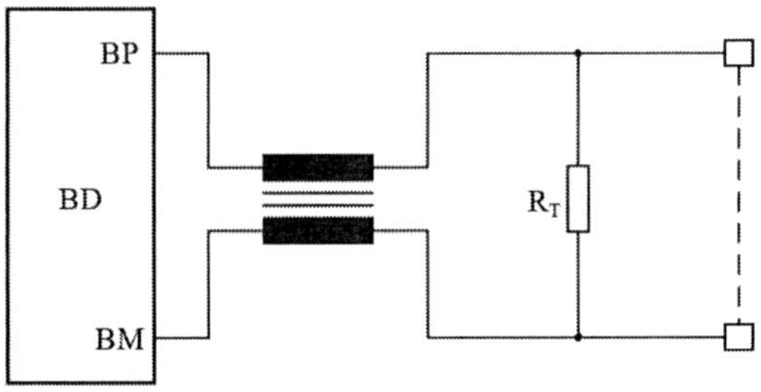

Figure 6. General Termination.

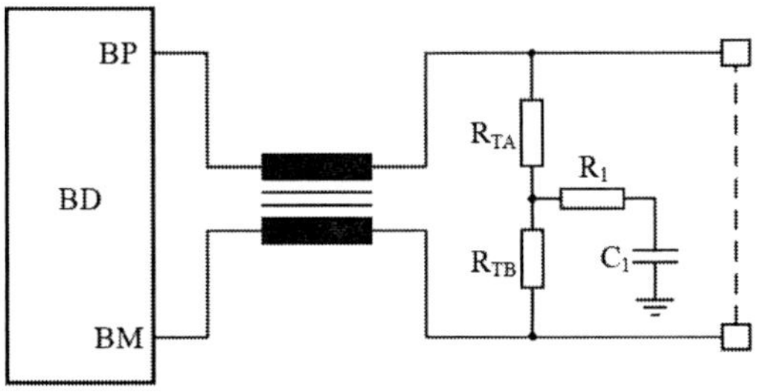

Figure 7. Separation Termination.

This design only requires the node to operate in the normal transceiver mode, then the TRXD0 and the TRXD1 are grounded. The pins 1-2 are used for power management and need to be used with the bus monitor. As there are no FlexRay bus monitor products currently on the market, these two pins are left empty. The pins 5-7 are the interface between TJA1080 and the communication controller for sending and receiving data. The pins 12-13 are used to feed status information back to the MCU. The BP and BM are a pair of differential signal lines, which is similar to CANH and CANL in the CAN bus.

In general, the simplest method is to terminate a resistor between BP and BM of the bus as shown in Fig.6. In order to better EMC performance, this design adopts the separation termination method, and places the common mode choke between the bus driver and the separation termination resistors as shown in the Fig.7. It divides the resistor in Fig.6 into two equal parts R_{TA} and R_{TB} , R_1-C_1 series branch is connected between the ground and the common node of the two separate termination resistors, providing a path for the common mode signal. For R_{TA} and R_{TB}, the resistance is as close as possible, and the equal value is put to use universally. The common mode choke has a large impedance to the common mode signal, which can make the current drive capability on the two signal cables consistent, but in the opposite direction.

In addition, it also includes power supply module, serial port module, etc. The main purpose is to make the FlexRay node working properly and easy to debug, and here is not described in detail.

5 Conclusion

According to the communication transmission demand of flight control system of pint and medium-sized UAV, this paper designs a FlexRay communication node based on the bus driver TJA1080 as the medium, which can realize the data exchange and sharing in the FlexRay bus. In this bus node design process, it focuses on a number of key points to ensure the real-time and reliability of communication transmission, and it still needs further investigations of FlexRay bus technology in aerospace applications.

References

1. J Lu, Research on the development of airborne bus technology, *Electronic Test*, **7**:72-73(2017).
2. Y Ding, Review of Airborne Avionics Bus System Development, *Avionics Technology*, **02**:1-7(2003).
3. X Ning, Airborne computer and its bus, National Conference on Signal and Intelligent Information Processing and Application, 54-55(2011).
4. L Li, J Zhang, Comparison of Several Common Buses in Airborne Computers, *Aeronautical Computing Technique*, **35**:103-105(2005).
5. Y Feng, H Dou, Application of CAN Bus in Aviation Field, *Advances in Aeronautical Science and Engineering*, **02**, 231-235(2011).
6. Z Gu, G Han, H Zeng, Q Zhao, Security-Aware Mapping and Scheduling with Hardware Co-Processors for FlexRay-based Distributed Embedded Systems, IEEE Transactions on Parallel & Distributed Systems, (2016).
7. Kanajan S A J. Sensitivity Analysis on Flexray Dynamic Segment Design Parameters_ International Conference on Systems & Networks Communications: International Conference on Systems & Networks Communications, (2009).
8. Y Li, X Chen, C Li, Design and Research of Serial Interface Unit Based on FlexRay Bus Flight Control Computer, *Computer Measurement & Control*, **02**:94-98(2016).
9. Y Liu, D Cao, X Lv, The flight control computer design of the FlexRay bus, *Microcontroller & Embedded Systems*, **04**:8-11(2015).
10. Z Liao, Q Gao, D Yuan, Design of FlexRay Node Drive Circuit Based on TJA1080, *Journal of Ordnance Equipment Engineering*, **10**:99-102(2015).

Optimized Puncturing of the Check Matrix for Rate-Compatible LDPC Codes

Liang Zhang, Jiahui Meng, Danfeng Zhao

College of Information and Communication Engineering, Harbin Engineering University, Harbin 150001, China

Abstract. In the wireless communication system, it is necessary to change the transmitted code word or code rate in real time according to the channel environment, designing and implementing rate compatible codes is of great practical significance. In allusion to the defects of algebraically constructed QC-LDPC codes algorithm and the time-varying characteristics of wireless channel, we propose an algorithm based on check matrix optimization puncturing technique to construct QC-LDPC codes in this paper. The algorithm selects a high rate matrix with excellent performance as the parent matrix, and then uses the backward node puncturing method to puncture the information bits in the code word from the last column of the check matrix by a multiple of the number of rows to obtain a low rate matrix. The simulation results show that the rate-compatible non-binary LDPC codes constructed by this algorithm not only reduces the complexity of hardware implementation, but also improves the error correction performance of the non-binary LDPC codes, realizes the rate-compatible coding at the same time, and improves the average effectiveness and reliability of the communication system.

1 Introduction

Low density parity check code (LDPC) [1-3], which is a good code for error correction performance close to Shannon limit, has become one of the research hot spots in recent years. The Quasi-Cyclic LDPC (QC-LDPC) code [4,5] is a very important type of LDPC code. Compared with a general LDPC code, it has a simpler coding and decoding complexity and it is very suitable for practical applications and realizations. Furthermore, the excellent performance can be achieved through the design of the structuring and degree distribution. At present, it has become a standard code word for many systems, such as deep space communication [6], DVB-S2 [7], WIMAX [8]. In wireless communication applications, the channel state changes continuously with time. In order to maximize the wireless communication throughput, it is necessary to change the code rate according to the channel state. Therefore, it is practically meaningful to implement a rate-compatible puncture code design. Puncture techniques are widely used in different types of forward error correction codes [9,10]. The rate-compatible code is a code containing a series of code rates, where the low code rate is included in the code of the high code rate [11]. In 2002, Li and Narayanan proposed a rate-compatible code based on LDPC, also known as rate-compatible LDPC (RC-LDPC). The puncturing technique was introduced to generate RC-LDPC codes by which a series of low rate codes can be obtained directly from a high code rate codes.

In this paper, according to the defects of QC-LDPC codes constructed by traditional algebra and the time-varying characteristics of wireless channels, an algorithm based on check matrix optimized puncturing technique to construct QC-LDPC codes (OPPMRC-LDPC) is proposed. The algorithm firstly selects a high code rate parent matrix with excellent performance, then uses the method of backward node puncturing method to delete the column from the end of the check matrix. The information bits in the codeword are punctured to obtain a low code rate codeword to achieve rate compatibility. By designing a set of check matrix that can be compatible with a variety of code rate QC-LDPC codes, the same codec and decoder can be used to complete different code rate switching, and each code rate has excellent error correction performance. The QC-LDPC codes constructed by this algorithm not only reduces the complexity compared with the traditional random structure method, but also does not generate the loop with a length of four. Then, through the design of the degree distribution, a lower error leveling layer code can be obtained, which not only reduces the hardware implementation complexity, but also improves the error correction algorithm, and at the same time impros the average effectiveness and reliability of the communication system. OPPMRC-LDPC has the same performance in each code rate as a single constructed LDPC code of the same code length with an optimal degree distribution.

2 Principle of algorithm

Designing an error correction coding system involves selecting a fixed code which has a certain code rate and error correction capabilities to adapt to the expected average or worst channel conditions. However, in many cases, it is more desirable to be flexible because the transmitted data has different protection requirements, and the channel is time-varying or its parameters are uncertain. Therefore, flexible channel coding and adaptive decoders are needed. This flexible system can be achieved by using a set of decoders, but in this case the hardware implementation complexity will increase significantly. Therefore, rate-compatible codes are introduced in systems with stringent hardware complexity requirements. In order to improve the transmission efficiency, it is desirable to transmit the check information as little as possible. On this foundation, we propose the OPPMRC-LDPC algorithm.

The design steps of specific error correction coding system are as follows:

1. Using PEG algorithm to generate basic matrix

Given the number of symbol nodes, check nodes, and symbol node degrees in the Tanner graph to be constructed, the edges between the symbol nodes and the check nodes are placed one by one so that each newly added edge affects the current graph as little as possible. That is, each non-zero element on the GF(q) field newly added to the check matrix H affects the girth of the current graph as little as possible.

2. Using QC-LDPC algorithm to generate a check matrix

(1) According to a specific degree distribution, a PEG algorithm is used to obtain a $J \times L$ dimensional binary basic matrix $\boldsymbol{H}_m = [m_{j,l}]$;

(2) Following the sequential trial method to determine the cyclic shift coefficient matrix $S_{J\times L}$;

A. If the number of row j and column l in H_m is "0", then the corresponding position shift coefficient $s_{j,l}$=Inf;

B. If the number of row j and column l in H_m is "1", then any number of $s_{j,l}$ in $\{0,1,\cdots,p-1\}$ will be selected. According to the determined loop length, using equation (1) to determined whether $s_{j,l}$ meets the requirements of the constraint type, if it meets the requirement, the number of $s_{j,l}$ is determined; if it does not meet the requirement, the new number is determined from B again;

$$\sum_{k=1}^{m}\left(s_{j_k,l_k} - s_{j_{k+1},l_k}\right) \neq 0 \bmod p \qquad (1)$$

Among them, $2 \le m \le i$, $1 \le j_k \le J$, $1 \le j_{k+1} \le J$, $1 \le l_k \le L$, $j_0 = j_m$.

C. The "1" in the basic matrix of the selected coefficient is replaced with a randomly selected number in $1,2,\cdots,q-1$ to generate a non-binary matrix.

(3) The corresponding basic matrix $\boldsymbol{H}_m = [m_{j,l}]$ and its cyclic shift coefficient matrix $S_{J\times L}$ are obtained, and the zero matrix is used to replace the "0" element at the position of (m,n) in H_m. The "1" element at the position of (m,n) in H_m is replaced by a non-binary matrix right shifted by $s_{j,l}$, finally, we can obtain the required check matrix $H_{(p\times J)\times(p\times L)}$.

The selection of non-binary numbers in each matrix block and the determination of the shift factor of each matrix block must be based on the principle of reducing or eliminating short loops. If the constructed matrix does not meet the requirements, it is necessary to reselect the shift factor or non-binary domain number. Due to the randomness in choosing the shift factor and the non-binary field number, the short loop of the matrix needs to be finally counted. The matrix which has too many short loops are directly discarded, the high-performance check matrix is searched again. That is a search and comparison process, the process stops until a good matrix is found to meet the requirements.

3. A puncturing algorithm based on check matrix optimization is compatible with multiple code rates

The implementation process of the puncturing algorithm based on check matrix optimization is to first select a high-rate parent code with excellent performance generated by the QC-LDPC codes check matrix construction algorithm described above, and then use the backward node puncturing algorithm to puncture the information bits in the codeword. That is, starting from the last column of the check matrix, several columns in the check matrix (m is the number of check matrix rows) are deleted in multiples of m, so as to obtain the required codeword with a low code rate. In case of good channel conditions, a high code rate QC-LDPC codes may be used. Conversely, when the channel environment is poor, a low code rate QC-LDPC codes may be used, thereby the efficiency of the communication system is improved. Since the non-zero elements in the puncturing matrix are sparse and have a certain regularity, the hardware implementation complexity of a code rate-compatible non-binary LDPC code communication system can be greatly reduced.

Given the puncturing code rate r_m, Gaussian approximation-based density evolution algorithm can be used to obtain the degree distribution under this rate condition. According to the optimized degree distribution, a puncturing algorithm optimized by the check matrix is used to obtain the non-binary LDPC code word under the punctured code rate r_m. From the perspective of hardware realizability and real-time performance, this scheme has practical significance.

4. Coding using partial parallel QC-LDPC codes encoding algorithm

The sub-matrix of the QC-LDPC generator matrix has a cyclic shift form, so that only the first row in the sub-matrix is stored, and the remaining elements are cyclically shifted by the first row, thereby the amount of encoding operations and storage is reduced. So we can use partial parallel encoding method to increase coding efficiency and reduce complexity. The check matrix of the non-binary QC-LDPC codes consists of $a \times b$ cyclic matrices of size $q \times q$, whose structure is shown in equation (2). $H_{i,j}$ is a cyclic matrix or zero matrix of size $q \times q$, the code length is $n = b \times q$, and the check bit length is $m = a \times q$.

$$H = \begin{bmatrix} H_{1,1} & H_{1,2} \cdots H_{1,(b-a)} & H_{1,(b-a+1)} \cdots & H_{1,(b-1)} & H_{1,b} \\ H_{2,1} & H_{2,2} \cdots H_{2,(b-a)} & H_{2,(b-a+1)} \cdots & H_{2,(b-1)} & H_{2,b} \\ \vdots & & & \vdots & \\ H_{a,1} & H_{a,2} \cdots H_{a,(b-a)} & H_{a,(b-a+1)} \cdots & H_{a,(b-1)} & H_{a,b} \end{bmatrix} \quad (2)$$

The check matrix must be a full rank matrix. The check matrix can be represented as $H = \begin{bmatrix} M & D \end{bmatrix}$, among that, D is composed of $a \times a$ cyclic matrix or zero matrix. The structure of the generator matrix obtained by the non-binary QC-LDPC check matrix conversion is shown in equation (3). The generator matrix consists of $(b-a) \times b$ circular matrices of size $q \times q$, and I is of size $q \times q$. In the matrix, O is a zero matrix of size $q \times q$, and $G_{i,j}$ is a cyclic matrix of size $q \times q$.

$$G = \begin{bmatrix} G_1 \\ G_2 \\ \vdots \\ G_{b-a} \end{bmatrix} = \begin{bmatrix} I & O & \cdots O & G_{1,1} & G_{1,2} & \cdots G_{1,a} \\ O & I & \cdots O & G_{2,1} & G_{2,2} & \cdots G_{2,a} \\ \vdots & \vdots & \vdots & \vdots & \vdots & \vdots \\ O & O & \cdots I & G_{(b-a),1} & G_{(b-a),2} & \cdots G_{(b-a),a} \end{bmatrix} \quad (3)$$

The generator matrix can be represented as $G = \begin{bmatrix} U & Z \end{bmatrix}$, U is made up of $(b-a) \times (b-a)$ cyclic matrices of size $q \times q$, and Z is composed of $(b-a) \times a$ cyclic matrices of size $q \times q$. Because the generator matrix and check matrix are satisfied $H \times G^{\mathrm{T}} = 0$, you can get:

$$M \times U^{\mathrm{T}} + D \times Z^{\mathrm{T}} = 0 \quad (4)$$

Since H is a full rank matrix, we have:

$$Z^{\mathrm{T}} = -D \times M \times U^{\mathrm{T}} \quad (5)$$

Let $\mathbf{z}_i = [\mathbf{g}_{i,1}, \mathbf{g}_{i,2}, \cdots, \mathbf{g}_{i,a}]$ be the first row of $[\mathbf{G}_{i,1}, \mathbf{G}_{i,2}, \cdots, \mathbf{G}_{i,a}]$ in the Z matrix, and $\mathbf{g}_{i,j}$ be the generator of $\mathbf{G}_{i,j}$. Since the generator matrix G is also composed of a cyclic matrix, partial parallel encoding is used for encoding. The SSRAA unit with the number a is simultaneously calculated to obtain all the check bits $\mathbf{p} = [\mathbf{p}_1, \mathbf{p}_2, \cdots, \mathbf{p}_a]$. Figure 1 shows the partial parallel coding structure of SSRAA-based non-binary QC-LDPC codes. It makes only $q(b-a)$ clock cycles to calculate all a segment check bits, which is a times faster than the classic encoding method.

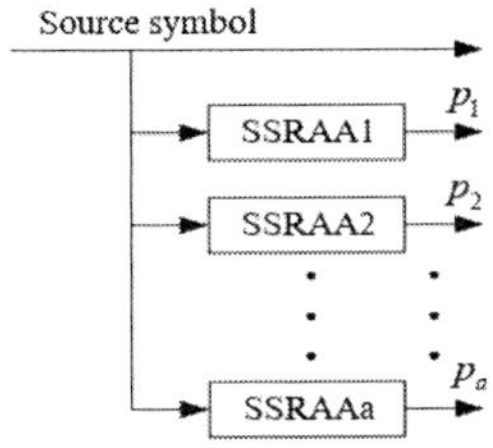

Figure 1. Partial parallel coding structure of non-binary QC-LDPC codes.

5. The software simulation part uses the Mixed Log-FFT-BP decoding algorithm to decode

In order to reduce the computational complexity, the Mixed Log-FFT-BP algorithm is used for decoding. The algorithm transforms the multiplication in the frequency domain into addition and lookup table operations in logarithmic domain, and then directly calculates the logarithm of the probability information to decode it in the logarithmic domain, which saves the process of solving the log-likelihood ratio, thereby computational complexity is reduced.

3 Simulation analysis

On the basis of QC-LDPC coding and Mixed Log-FFT-BP decoding, we analyzed the performance of the constructed check matrix. The parameters are set as follows: the number of multiple domains is 8, the code rate is 1/2, 2/3, 3/4, the maximum iteration number is 20 times, the fixed check bits number is 576bit, the channel is Gauss white noise channel, and BPSK modulation is used. The BER performance analysis with different rate is shown in Figure 2. We can see that when signal to noise ratio is 3dB, the bit error rate at 1/2 bit rate can reach 10^{-8} level, the bit error rate under 2/3 code rate can reach 10^{-7} level, and bit error rate under the 3/4 code rate is 10^{-6} level. Therefore, compared with reference [12,13], the check matrix constructed by the puncturing algorithm optimized by the check matrix improves the error correction performance of the non-binary LDPC code, and at the same time realizes the code rate-compatible coding.

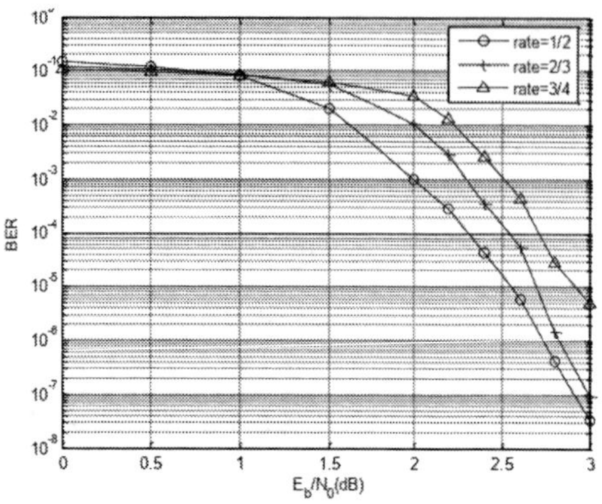

Figure 2. BER analysis of different code rates.

4 Hardware implementation

The hardware implementation of the encoder is based on FPGA, making full use of the parallel computing features of the FPGA to increase the encoding efficiency and reduce the encoding delay. The code rate-compatible non-binary LDPC coding design adopts partial parallel QC-LDPC codes coding algorithm and is divided into two main parts: the code rate control module and the core coding module. The structure of the design is shown in Figure 3.

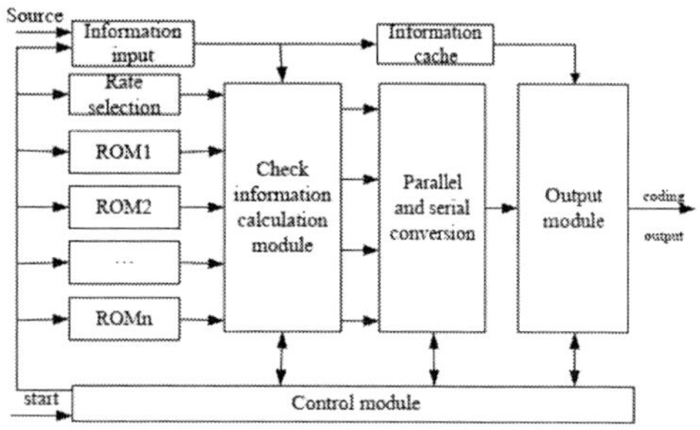

Figure 3. Structure diagram of code rate-compatible non-binary LDPC encoder.

Since the QC-LDPC matrix has a quasi-cyclic characteristic and made up of cyclic matrix blocks of $q\times q$, the control of rate can be achieved by controlling the number of sub matrices involved in operation. Each column block corresponds to q-bit check bits, so that all column blocks can be calculated synchronously, all check bits are obtained within the same clock, and coding efficiency is improved.

1. Rate control module

According to the characteristics of the sub-matrix as the unit of the generator matrix and the check matrix, the rate control can be achieved by controlling the number of sub-matrices participating in the operation. When we construct the check matrix, the low code rate is included in the high code rate, and the generator matrix obtained through the check matrix also has this feature. The code rate-compatible of the check matrix is represented by the fact that the matrix at low rate behaves as part of the high rate matrix row, while the generator matrix shows that the matrix at low rate is part of the high rate matrix column. Therefore, the length of the column affects the coding rate, and different matrix blocks with different lengths can be used to achieve code rate compatibility at different code rates.

2. Core coding module

The core coding module is the core part of the encode which completes the calculation of the check bit and outputs the code word. First, the control module controls the selection and displacement of generators. Each check information calculation module uses the generated generator and information bits to perform Galois field operations to obtain q-bit check bits, and all check information calculation modules simultaneously perform check bit calculation. Then all check bits blocks are converted to check bits in the codeword, and finally combined with the cache information sequence into a complete codeword.

5 Conclusion

In order to improve the adaptation performance of non-binary LDPC codes, we propose an RC-LDPC code design method based on check matrix puncturing optimization, which combines the rate-compatible conditions to optimize the progressive performance of parent codes. According to the desired code rate, the information bits that need to be deleted are determined, and the backward node puncturing algorithm performs puncturing from the back to the forward. The matrix constructed by this algorithm has no loop and has good performance. The simulation results show that the algorithm reduces the hardware implementation complexity and improves the error correction performance of the LDPC code.

References

1. S. Zhao, X. Ma, X. Zhang. A Class of Nonbinary LDPC Codes with Fast Encoding and Decoding Algorithms. *IEEE Transactions on Communications*, **61**, 1 (2013)
2. M. Gholami, M. Samadieh. Design of Binary and Nonbinary Codes from Lifting of Girth-8 Cycle Codes with Minimum Lengths. *IEEE Communications Letters*, **17**, 4 (2013)
3. D.J.C. MacKay. Good error-correcting codes based on very sparse matrices. *IEEE Transactions on Information Theory*, **45**, 2 (1999)
4. V. Dragoi, H.T. Kalachi. Cryptanalysis of a public key encryption scheme based on QC-LDPC and QC-MDPC codes. *IEEE Communications letters*, **2**, 22 (2018)
5. C.M. Stuart, P. Deepthi. Nonlinear cryptosystem based on QC-LDPC codes for enhanced security and reliability with low hardware complexity and reduced key size. *Wireless personal communications*, **3**, 96 (2017)
6. A.A.S. Afshar, T. Eghlidos, M.R. Aref. Efficient secure channel coding based on quasi-cyclic low-density parity-check codes. *IET communications*,

2, 3 (2009)

7. J.J. Zhang, M.K. Dong, D. Wang. Construction of Quasi-cyclic low-density parity-check codes for simplifying shuffle networks in layer decoder. *China Communications*, **12**, 10 (2013)
8. X.M. Wang, T.T. Ge, J. Li. Efficient multi-rate encoder of QC-LDPC codes based on FPGA for WIMAX standard. *Chinese journal of electronics*, **2**, 26 (2017)
9. X.L. Liu, Q.H. Liu, Z.Q. Chen. Study on construction methods of RC-LDPC code. *Journal of Guilin university of electronic technology*, **6**, 30 (2010)
10. Y. Xu, B. Liu, L. Gong. Improved shortening algorithm for irregular QC-LDPC codes using known bits. *IEEE transactions on consumer electronics*, **3**, 57 (2011)
11. K. Zhang, X.M. Huang, Z.F. Wang. A High-throughput LDPC decoder architecture with rate compatibility. *IEEE transactions on circuits and systems i-regular papers*, **4**, 58 (2011)
12. N.N. Tong, Y.P. Wu. A new nonbinary quasi-cyclic LDPC codes construction algorithm. *International journal of advancements in computing technology*, **5**, 8 (2013)
13. D.F. Zhao, N.N. Tong, Y.P. Wu. The research of construction algorithm of irregular QC-LDPC codes based on masking technique. *Journal of Jiamusi university*, **6**, 25 (2007)

Universal Impulse Noise Suppression Using Extended Efficient Nonparametric Switching Median Filter

M. H. Suid[1], M. A. Ahmad[1], M. I. F. M. Hanif[2], M. Z. Tumari[3] and M. S. Saealal[3]

[1]*Faculty of Electrical & Electronic Engineering, Universiti Malaysia Pahang, 26600 Pekan, Pahang, Malaysia.*
[2]*SEGi University, No. 9 Jalan Teknologi, Taman Sains Selangor, Kota Damansara, PJU 5, 47810 Petaing Jaya, Selangor.*
[3]*Faculty of Engineering Technology, Universiti Teknikal Malaysia Melaka, Hang Tuah Jaya, 76100 Durian Tunggal, Melaka, Malaysia.*

Abstract. This paper presents a filtering algorithm called extended efficient nonparametric switching median (EENPSM) filter. The proposed filter is composed of a nonparametric easy to implement impulse noise detector and a recursive pixel restoration technique. Initially, the impulse detector classifies any possible impulsive noise pixels. Subsequently, the filtering phase replaces the detected noise pixels. In addition, the filtering phase employs fuzzy reasoning to deal with uncertainties present in local information. Contrary to the existing conventional filters that only focus on a particular impulse noise model, the EENPSM filter is capable of filtering all kinds of impulse noise (i.e. the random-valued and/or fixed-valued impulse noise models). Extensive qualitative and quantitative evaluations have shown that the EENPSM method performs better than some of the existing methods by giving better filtering performance.

Keywords: Image processing, impulse noise, digital image, noise filtering, nonparametric switching median filter.

1 INTRODUCTION

With the usage of multimedia material becoming more widespread from day to day, visual information from high quality digital images plays an important role in many daily life applications. Unfortunately, digital images are frequently subjected to the contamination of impulse noise due to the interferences generated during transmission, acquisition in noisy environment [1]. Luckily, due to the advancement in digital technologies the level of noise density in digital images has dropped significantly to the low contamination rate. Still, even at low densities, the occurrence of impulse noise can severely damage the information in the original image. Therefore, it is essential to remove impulse noise effect before carrying any subsequent image processing task (e.g. segmentation, object recognition, data compression etc.). Mostly, these subsequent processing steps are largely affected by the quality of the filtered image [2].

For this purpose, many filters have been proposed. The median filter for instance, is a well-known nonlinear filter for suppressing impulsive noise due to its effectiveness and high computational efficiency [3],[4]. Despite its effectiveness in smoothing noise, the median filter tends to blur fine details and often destroys edges due to its clumsy filtering property that treats all the pixels equally without considering whether or not it is noise-free pixel.

The aforementioned drawbacks have led to the development of various switching-based filters, e.g. the switching median (SWM) filters along with the centre weighted switching median (CWSWM) filter [5], the Laplacian switching median (LSM) filter [6], progressive switching median (PSM) filter [7] and the multi-state median (MSM) filter [8], etc. Basically, this class of filtering scheme works based on the impulse detection mechanism which uses a fixed size filtering window and predefined threshold value to differentiate between noise and noise-free pixels. With the noise detector, these filters are shown to be more effective in terms of the detail and edge preservation compared to the uniformly applied conventional median filters. However, one disadvantage is that the switching rule is typically based on a fixed threshold for locally obtained statistics. This approach in certain circumstances tends to yield problem of pixel's misclassification and fails to replace the noise pixels.

Of late, works in [9] has come out with a more flexible switching-based filter called efficient nonparametric switching median (ENPSM) filter, for detail-preserving restoration. The ENPSM filter is customary based on the combination of local variance threshold in the impulse noise detection module and recursive restoration technique in the pixel restoration module. Even though this method performs well, yet it only touched on the filtering of the random-valued (RV) impulse noise corrupted cases.

Thus, in this paper we take one step further by focusing on the detection and suppression of any type of impulse noise models. By using the same existing recursive filtering technique in the ENPSM filter, the proposed technique called extended efficient nonparametric switching median (EENPSM) filter is equipped with histogram-based impulse noise detector which is specifically designed for accurate fixed-valued (FV) impulse noise detection.

2 IMPULSE NOISE MODEL

In theory, impulse noise contamination amplitude could fall either within the image dynamic range (i.e. RV impulse) or out of the range (i.e. FV impulse) and usually only certain percentages of pixels are altered. For more detail, let $x(i, j)$ and $o(i, j)$ be the gray level of the noisy image and the original image at location (i, j), respectively. Then, the impulse noise model with noise density ρ can be defined as:

$$x_{(i,j)} = \begin{cases} n_{(i,j)} : & \text{with probability } \rho \\ o_{(i,j)} : & \text{with probability } 1 - \rho \end{cases} \tag{1}$$

where $n(i, j)$ is the noise pixel value. The image is likely to be contaminated by the RV impulse noise when $n(i, j)$ is uniformly distributed within the image dynamic range, i.e. $n(i, j) \in [Nmin, Nmax]$. On the other hand, the simplest impulse noise model is the FV impulse noise; where $n(i, j)$ is assumed to take the maximal and minimal intensities, i.e. $n(i, j) \in (Nmin, Nmax)$.

Although many researchers have paid special attention in obtaining a good filter for on a particular impulse noise model, this proposed technique takes one step further by focusing on the removal of RV and FV impulse noise all at once. In practical, identifying this scenario is more challenging compared to smoothing the RV noise or FV noise alone since it is closer to the real-world situations.

3 EFFICIENT NONPARAMETRIC SWITCHING MEDIAN FILTER

In this section, the conventional ENPSM filtering algorithm proposed in [9] will be briefly reviewed. Given a noisy image and initial filtering window $W(i, j)$ of size $(2N+1)\times(2N+1)$, the RV noise detection stages of ENPSM are described as follows:

<u>Step 1</u>: Sort all elements within $W(i, j)$ in ascending order and find the median pixel $m(i, j)$.

<u>Step 2</u>: Compute the absolute luminance differences $d(i\pm k, j\pm l)$ between $m(i, j)$ and all pixels in $W(i, j)$;
$d(i\pm k, j\pm l) = |x(i\pm k, j\pm l) - m(i,j)|$; for $-N \le k,l \le N$

<u>Step 3</u>: Rearrange each value obtained in $d(i\pm k, j\pm l)$ and set the predefined threshold T_{ENPSM} as; $T_{ENPSM} = \text{med}\{d(i\pm k, j\pm l) \mid : -N \le k,l \le N\}$.

<u>Step 4:</u> Mark the locations of 'noise' pixels (i.e. $d(i\pm k, j\pm l) > T_{ENPSM}$) and 'noise-free'(i.e. $d(i\pm k, j\pm l) \le T_{ENPSM}$) pixels in the binary noise map $M(i, j)$. Slide $W(i, j)$ to the next pixel and repeat Step 1 to Step 4 until the process is completed for the entire image.

This first part of this detection process is sufficient to handle impulse noise density (up to 20%), largely for RV impulse noise. However, it is still not appropriate in the ENPSM to detect and replace the FV impulse noise accurately, particularly when the noise density is moderate or high.

4 EXTENDED EFFICIENT NONPARAMETRIC SWITCHING MEDIAN FILTER

The proposed filter is a modification of the original ENPSM filter, which operates on the same sliding window spatial filter that targets each pixel in a filtered image sequentially. The difference of the proposed filter with the former ones is that the proposed EENPSM filter modifies the existing ENPSM by adding one more process, in case when the case of FV contamination happens.

4.1 Noise detection stage

Taking the FV impulse noise into account, the noisy image histogram will be utilized by the proposed EENPSM filter. It is known that the peak intensities at the ends normally represent the FV noise intensities; e.g, see [10]. By employing the local maximum, the two fixed-valued impulsive intensities can be found by traversing the noisy image histogram from both ends and directed towards the centre of the histogram simultaneously. Once the local maximums denoted as L_{Min} and L_{Max}, are found then the search will be stopped immediately. The detected local maximums represent the two fixed-valued FV impulse noise intensities. Furthermore, FV noise model is constructed based on the assumption that noise pixels will assume the two extreme values in the image dynamic range. Under some real-world situations, these FV noise pixels can be replaced by close approximations of their actual noise intensities [11-13]. For example, pixels with the intensities value of 0 are possible to be replaced with 1 or 2, and intensity value 255 will be replaced with 254 or 253 in an image stored as an 8-bit integer. Hence, distortion will be hardly detected by common FV impulse noise filters. At the end of the detection stage, a two-dimensional binary noise detection map $M(i, j)$ is formed based on:

$$M(i,j) = \begin{cases} 1, & \left[d(i\pm k, j\pm l) > T_{ENPSM}\right] \\ & \cap\left[x(i,j) = L_{Min} \cap L_{Max}\right] \\ 0, & d(i\pm k, j\pm l) \le T_{ENPSM} \end{cases} \tag{2}$$

where $M(i, j)$ in Step 4 is modified to produce a new version of $M(i, j)$ as shown in equation (2). Logic '1s' in the equation shows the positions of noisy pixels and logic '0s' intended for those non-noisy ones.

4.2 Noise filtering stage

After binary noise mask $M(i, j)$ is formed, those pixels marked with $M(i, j) = 1$ is then will be swapped by the estimated median value. Otherwise, the filtering action is skipped when $M(i, j) = 0$ and the pixels will be left unprocessed. Once again, the proposed EENPSM algorithm uses a square filtering window $W_{filter}(i, j)$ with odd $(2N+1)\times(2N+1)$ dimensions and it is given as:

$$W_{filter}(i,j)=\{x(i\pm k,j\pm l),\ldots,x(i,j),\ldots,x(i\pm k,j\pm l)\};\quad \text{where} -N_{filter}\le k,l\le N_{filter} \tag{3}$$

For every '1s' pixel detected, the estimated median value is counted using only 'noise-free' pixels in the current filtering window. The calculation process is carried out using:

$$m_{est}(i,j)=\text{med}\{x(i\pm k,j\pm l),..,x(i,j),..,x(i\pm k,j\pm l)\};\quad \text{with } M(i\pm k,j\pm l)=0 \tag{4}$$

This criterion of choosing only 'noise-free' pixels is imposed to avoid selecting a 'noise' pixel as the estimated median pixel. Finally, the correction term to restore a detected noisy pixel is a linear combination between the current processing pixel and the estimated median pixel. The restoration term is given here as;

$$y_{EENPSM}(i,j)=\left[1-M(i,j)\right]\cdot x(i,j)+M(i,j)\cdot m_{est}(i,j) \tag{5}$$

5 SIMULATION RESULTS AND DISCUSSIONS

In this section, the practicability of the proposed EENPSM filter will be compared to the original ENPSM filter based on their simulation results. Three examples are provided to verify and justify the ideas described in Section 4. In this experiment, the original 512×512, 8 bits gray scale images *Flower*, *Pens* and *Yacht* were used in the simulation of the implemented filters. Those images are commonly used in image processing research and studies. Each of them was corrupted with fixed-valued and random-valued impulse noise ranging from 10% to 30%. Both qualitative and quantitative assessments are employed to assess the performances of the proposed EENPSM and the original ENPSM filter. The quantitative assessment used here is the peak signal to noise ratio (PSNR) which is defined as:

$$PSNR = 10\log_{10}\left(\frac{255^2}{MSE}\right)dB \tag{6}$$

where MSE is the mean-squared error given as:

$$MSE=\left(\frac{1}{M\times N}\right)\sum_{i=0}^{M-1}\sum_{j=0}^{N-1}\left[o(i,j)-y(i,j)\right]^2 \tag{7}$$

For the above mentioned formula, $M\times N$ is the image size with 'M' rows and 'N' columns, $y(i, j)$ is the filtered image and $o(i, j)$ is the original noise-free image.

As can be seen in Figure 1, at 10% impulse noise density, the noise filtering performance of EENPSM filter is basically similar to the original ENPSM filtering algorithm. Both filters are found to be able of producing visible restored image at this noise level since the density of impulse noise is still low and not yet form any noise patches at this level

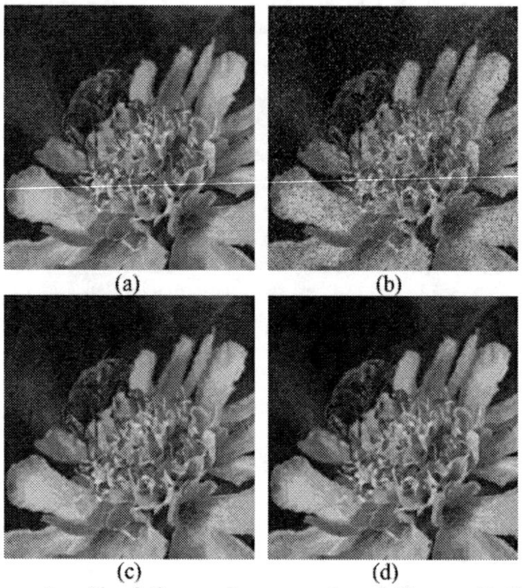

Figure 1. Simulation results on a portion of *Flower* with 10% density of impulse noise using; (a) original image, (b) noisy image, (c) ENPSM and (d) EENPSM.

However, in the *Pens* image (see Figure 2), we may notice that the proposed EENPSM filter gives better and clearer filtering result compared to the original ENPSM algorithm. The noise particles and effects are significantly reduced and at the same time the image details are well preserved. In contrast, we may be able to notice that some small noise spots are remained intact on the resultant images produced by the ENPSM filter. It is found that the original ENPSM filter has problem in the case of FV noise blotches (a place in image where a large number of fixed-valued impulse pixels may connect).

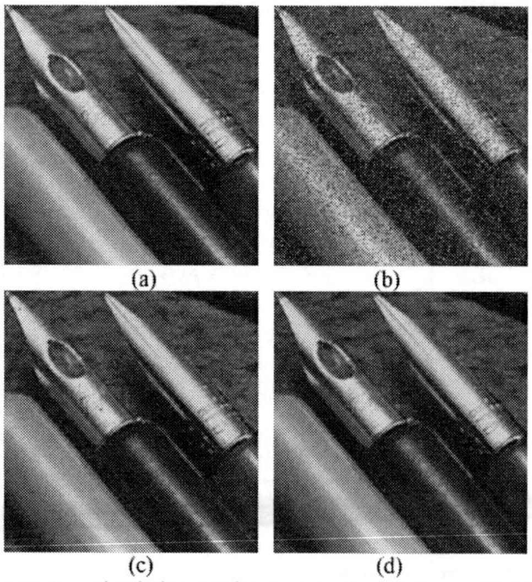

Figure 2. Simulation results on a portion of *Pens* with 20% density of impulse noise using; (a) original image, (b) noisy image, (c) ENPSM and (d) EENPSM.

The similar phenomenons are obtained for the *Yacht* test image (shown in Figure 3), where the proposed EENPSM filtering algorithm consistantly outperforms the ENPSM filter by giving clearer image; even the density of noise in this image is increasing. This is due to the ability of the proposed algorithm to distinguish the FV noise pixels more dexterously as compared to the previous filtering version.

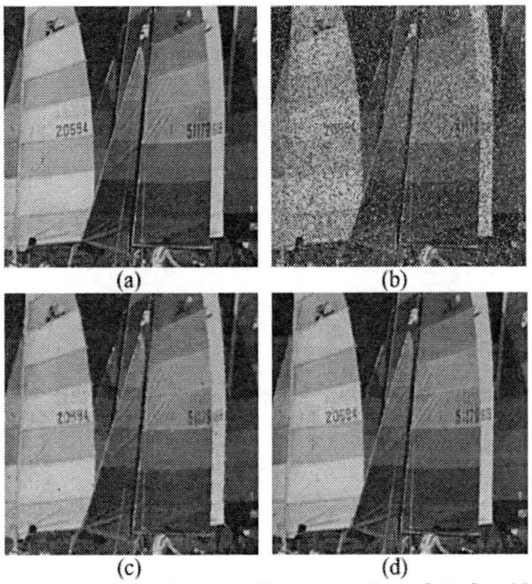

Figure 3. Simulation results on a portion of *Yacht* with 30% density of impulse noise using; (a) original image, (b) noisy image, (c) ENPSM and (d) EENPSM.

Table 1. Comparison of PSNR on Different Noise Level Restoration for *Flower* (Test Image)

Algorithms	PSNR (dB)		
	10%	20%	30%
ENPSM	33.9590	31.9943	30.6433
Proposed EENPSM	**34.5308**	**33.2594**	**31.6185**

Table 2. Comparison of PSNR on Different Noise Level Restoration for *Pens* (Test Image)

Algorithms	PSNR (dB)		
	10%	20%	30%
ENPSM	36.1737	33.5591	30.6339
Proposed EENPSM	**37.5011**	**34.8123**	**32.845**

Table 3. Comparison of PSNR on Different Noise Level Restoration for *Yacht* (Test Image)

Algorithms	PSNR (dB)		
	10%	20%	30%
ENPSM	31.5160	29.1585	26.6862
Proposed EENPSM	**33.3688**	**30.3931**	**28.6231**

Meanwhile, the numerical results for the used standard test images (i.e *Flower*, *Pens* and *Yacht*) are presented in Tables 1-3, respectively. Overall, it is made known that the proposed EENPSM unfailingly outperforms the previous algorithm at all level of universal impuse contamination case. It is evident that EENPSM filtering performance is tremendously consistent. In contrast, inconsistentce performances have been shown by the ENPSM filter with their PSNR values to have decreased intensely especially at the 30% noise level. Indirectly this result also shows that the conventional methods were unable to cater for the occurrences of noise in a proper manner especially when it comes to the fixed-valued contamination cases.

6 CONCLUSIONS

In this article, an extended version of ENPSM filter namely EENPSM for effective universal impulse noise restoration is presented. The variance thresholding and recursive pixel restoration techniques that involved in the design of the filter make it able to suppress both random-valued and fixed-valued impulse noise effectively, at the same time preserving fine image edges and textures. Furthermore, this filter does not require any special tuning of parameter since its predefined threshold is established based on nonparametric framework. The simulation results indicate that a better noise filtering performance is achieved. In addition, the fuzzy reasoning could be embedded as part of its filtering mechanism, which permits us to exploit the effectiveness of fuzzy paradigm in handling imprecise local information. Overall, it is a feasible approach for removing the effects of low level universal impulse noise in digital images.

Acknowledgement

This work was partially supported by the Research and Innovation Department, University Malaysia Pahang and the Ministry of Higher Education under research grant RDU1703151.

References

1. X. Lan, Z. Zuo, Random-valued impulse noise removal by the adaptive switching median detects and detail preserving regularization, (2013).
2. K. K. V. Toh, H. Ibrahim and M. N. Mahyuddin. Cluster-based Adaptive Fuzzy Switching Median Filter for Universal Impulse Noise Reduction. *IEEE trans. Consumer Electron.*, **56(4)**, 2560-2568 (2010).
3. J. Astola, P. Kuosmanen, *Fundamentals of Nonlinear Digital Filtering*, CRC, Boca Raton, FL, (1997).
4. R.C. Gonzalez, R.E. Woods, Digital Image Processing, Prentice Hall, New Jersey, (2002).
5. T. Sun and Y. Neuvo. Detail-Preserving Median Based Filters in Image Processing. *Pattern Recognition Letters*,**15** (1994) 341–347.

6. S. Zhang and M. A. Karim. A new impulse detector for switching median filters. *IEEE Signal Processing Letters*, **9 (11)**, 360-363, (2002).
7. Z. Wang and D. Zhang. Progressive switching median filter for the removal of impulse noise from highly corrupted images. *IEEE Trans. on Circuits and Systems II*, **46 (1)**,78-80, (1999).
8. T. Chen and H. R. Wu. Adaptive impulse detection using centre-weighted median filter. *IEEE Signal Processing Letters*, **8 (1)**, 1-3 (2001).
9. M. H. Suid and N. A. M. Isa. Random valued impulse noise reduction in digital image using efficient nonparametric switching median filter. *Scientific Research and Essays* vol. **7(7)**, 813-823, (2012).
10. K. K. V. Toh and N. A. M. Isa "Noise Adaptive Fuzzy Switching Median Filter for Salt-and-Pepper Noise Reduction", *IEEE Trans. Consumer Electron.*, vol. **17(3)**, 281–284, (2010).
11. N. I. Petrovic and V. Crnojevic, " Universal impulse noise filter based on genetic programming," *IEEE Trans. Image Process.*, vol. **(17)(7)**, 1109-1120, (2008).
12. M. Ismaeil, K. Pritamdas, K. J. K. Devi and S. Goyal, "Performance analysis of new adaptive decision based median filter on FPGA for impulsive noise filtering," *1st International Conference on Electronics, Materials Engineering and Nano-Technology (IEMENTech)*, Kolkata, 2017, pp. 1-5, (2017).
13. M. S. Darus, S. N. Sulaiman, I. S. Isa, Z. Hussain, N. M. Tahir and N. A. M. Isa, "Modified hybrid median filter for removal of low density random-valued impulse noise in images," *6th IEEE International Conference on Control System, Computing and Engineering (ICCSCE)*, Batu Ferringhi, 2016, pp. 528-533, (2016).

End-to-End Mandarin Recognition based on Convolution Input

WANG Yanzhe , ZHANG LiMin, ZHANG Bingqiang, and Li Zhenyu

Institute of Information Fusion, Naval Aviation University, 264000 Yantai Shandong, China

Abstract. The cross-entropy criterion of mainstream neural network training is to classify and optimize each frame of acoustic data, while the continuous speech recognition uses the sequence-level transcription accuracy as a performance measure. In view of this difference, an end-to-end speech recognition system based on sequence level transcription is constructed in this paper. The model uses convolution neural network to deal with the input features, selects the best network structure, and performs two-dimensional convolution in the time and frequency domains. At the same time, neural network uses batch normalization technology to reduce generalization error and speed up training. Finally, the hyper-parameters in decoding process are optimized to improve the modelling effect. Experimental results show that the system performance is improved a lot, better than mainstream speech recognition systems.

1 Introduction

It is a complex task to construct modern automatic speech recognition system, which is based on strict design process, including input feature, acoustic model, language model and hidden Markov model. The long-term goal of ASR is to be able to deal with people's speech in different environments effectively, which is quite challenging for the traditional GMM-HMM hybrid model. A good acoustic model should be capable of simulating various acoustical changes in speech signals effectively in order to obtain robustness under different speech and environmental conditions.

In recent years, acoustic models based on deep neural networks (DNN) and convolutional neural networks (CNN) have achieved good results in some challenging large vocabulary speech recognition tasks [1][2].

However, DNNs have a disadvantage in that they are completely connected and cannot fully utilize the structural locality in the feature space. Similar to model objects in computer vision, acoustic signals may contain basic two-dimensional feature patterns. CNN can better explain small changes and disturbances in the feature space that may be caused by different speaking styles and environments.

In speech recognition, hybrid neural networks are trained to use cross-entropy criteria [2] as objective functions to classify individual frames of acoustic data, which is very different from the accuracy of transcription at the sequence level. This hybrid approach maybe suboptimal for solving the final task because different modules are trained with different standards so that extra hyperparametric adjustments are maybe required for each training stage.

Graves proposed a combination of deep, Long Short-Term Memory (LSTM) networks [3] and linking Connectionist Temporal Classification (CTC) objective functions [4] in [5], which can directly transcribe audio data into text without the need of inter-mediate voice representation. Moreover, there is no need for any pre-alignment between the input sequence and the target sequence, so end-to-end speech recognition represents the trend of future development.

A lot of work has been done in English speech recognition before, so this paper combines deep bidirectional Long Short-Term Memory (DBLSTM) networks with CTC to establish an end-to-end speech recognition model for Mandarin. Model input uses CNN, and neural networks use batch normalization (BN) technique. According to the phonetic features of Chinese Mandarin, characters are used as modelling units, combined with dictionary and speech models when decoding. Experimental results show that the recognition of Chinese Mandarin in resource-constrained conditions results in a good recognition rate.

2 End-to-end speech recognition model

Applying CNN in ASR can reduce spectrum changes and effectively model the spectral correlation in ASR acoustic features. Applying LSTM can learn the dependence information between long-term sequences. Finally, it can be combined with CTC to build a true end-to-end. The speech recognition model (Fig. 1) replaces the traditional method of relying on the HMM model.

2.1 LSTMP

The speech signal has the characteristics of long-range correlation. Standard feedforward networks typically only handle information in the frame's fixed-length sliding window, while LSTM can encode sequence history as their internal state, and can predict phonemes based on all the speech features observed prior to the current frame.

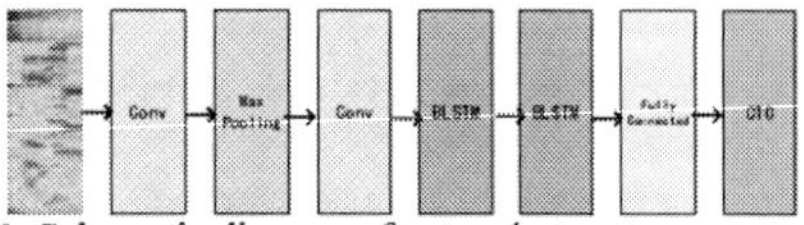

Figure 1. Schematic diagram of network structure

Because speech signal depends heavily on the front and back, the output is usually based on previous and subsequent states. So BLSTM should be used to solve the problem of using the future information of speech.

In order to solve the problem of training a DBLSTM that is slow and difficult to fit. H.Sak [6] combined the matrix decomposition with LSTM to propose a long and short memory network structure with a projection layer. A single projection vector with a lower size linear projection layer than the LSTM output is set behind the LSTM layer.

$$i_t = \sigma(W_{ix}x_t + W_{ip}p_{t-1} + W_{ic}c_{t-1} + b_i) \quad (1)$$

$$f_t = \sigma(W_{fx}x_t + W_{fp}p_{t-1} + W_{fc}c_{t-1} + b_f) \quad (2)$$

$$a_t = g(W_{cx}x_t + W_{cp}p_{t-1} + b_c) \quad (3)$$

$$c_t = f_t c_{t-1} + i_t a_t \quad (4)$$

$$o_t = \sigma(W_{ox}x_t + W_{op}p_{t-1} + W_{oc}c_t + b_0) \quad (5)$$

$$p_t = W_{pm}(o_t h(c_t)) \quad (6)$$

$$y_t = soft\max(W_{yp}p_t + b_y) \quad (7)$$

i , f , o , a and c represents the input gate, the forget gate, the output gate, the unit input activation and the unit state vector, respectively. W and b represents the weight matrix and the offset vector, respectively. W_{ic} , W_{fc} and W_{oc} is the diagonal weight matrix for each door and peephole connection.

2.2 CNN

Recently, CNN has been introduced into ASR. Because of its powerful modelling performance of sound signal, it has better results than DNN in many tasks [2].

CNN usually consists of one or more sets of convolutional layers and sampling layers. Each convolutional layer has a number of convolutional filters, which are convolved with lower activations to learn different local features in the speech signal. Then the maximum sampling method (Max-pooling) is generally used to down-sample the output activation at a fixed length, thus reducing the parameters and retaining most of the feature information to improve the performance.

In computer vision, when the convolutional layer convolves each image, the filters' weights are all shared. However, for speech signals, the local structures appearing in different frequency bands may behave in completely different ways, so the idea of limited-weight sharing of convolutional layers is introduced into ASR.

2.2.1 Full Weight Sharing

Assuming the input of Neural Network is $x \in R^{A \times B}$, where A is the number of features representing an input frequency band and B is the number of the input frequency bands. Let's assume that $x = [x_1\ x_2\ ...\ x_B]$, where x_b is the feature vector representing band b. The activations of the convolution layer can be computed as:

$$h_{j,k} = \theta(\sum_{b=1}^{s} w_{b,j}^T x_{b+k-1} + a_j) \quad (8)$$

where $h_{j,k}$ is the convolution layer's output of the jth feature map on the kth convolution layer band, s is the filter size, $w_{b,j}$ is a weight vector representing the bth band of the jth filter, a_j is the bias of the jth feature map. After the input and filters are weighted average, the output node value of the network is calculated from the nonlinear function θ .

The pooling function, which computes some statistics of the activations, is typically applied to the neurons along a window of frequency bands and generated from the same feature map in the convolution layer. Max pooling function simply computes the maximum value of the feature over the corresponding frequency bands. The max-pooling activations can be computed as :

$$p_{j,m} = \max_{k=1}^{r}(h_{j,(m-1)\times n+k}) \quad (9)$$

where $p_{j,m}$ is the pooling layer's output of the jth feature map and mth pooling layer band, n is the sub sampling factor, and r is the pooling size, which is the number of bands to be pooled together.

2.2.1 Limited Weight Sharing

Unlike FWS, LWS uses multiple different filters to convolve different frequency windows of the speech signal. The weight for input is not fully shared. The neurons in the pooling layer represent the sampling of specific feature maps obtained by different filters convolutions. It is as if the convolution and pooling layers are divided into many sections, each of which processes only a limited range of the input bands and generates only one output band in the pooling layer. Convolution and pooling are calculated as follows:

$$h_{j,k}^{(m)} = \theta(\sum_{b=1}^{s} w_{b,j}^{(m)} x_{(m-1)\times n+b+k-1}^T + a_j^{(m)}) \quad (10)$$

$$p_{j,m} = \max_{k=1}^{r}(h_{j,k}^{(m)}) \quad (11)$$

where $h_{j,k}^{(m)}$ is the value of the kth band of the jth feature map at the mth convolution layer section, and $p_{j,m}$ is the value of the jth feature of the mth pooling layer band.

2.3 Connectionist Temporal Classification

Traditional hybrid neural networks are suboptimal for ASR, because training using different standards at different stages may not be optimal for solving the final task. CTC can directly map speech features to labels without relying on the alignment between audio sequences and label sequences. The output of the softmax layer of RNN is generally used as the input of the CTC. An extra label is set to represent the blank, which is used to estimate the probability of not outputting the label at a specific moment, for a total of $K+1$ labels.

Given an input sequence X whose length is T, the probability of the softmax layer outputs a label or blank index k at time t is:

$$P(k \mid t, x) = \frac{\exp(y_t^k)}{\sum_{k'} \exp(y_t^{k'})} \quad (12)$$

$p(z^l \mid x)$ is the product of the output probability of each time step:

$$p(z^l \mid x) = \prod_{t=1}^{T} P(k \mid t, x) \quad (13)$$

z^l denotes all possible alignment of x and l in grid coding. Alignment includes label repetition and blank label insertion. There are several kinds of paths corresponding to an audio sequence, so we must first remove the repeated tags and blanks in the path, and map these paths to transcription. The label sequence l is represented by a set of all possible CTC paths mapped to l, and the probability is the sum of all path probabilities:

$$p(l \mid x) = \sum_{p \in \phi(l)} p(z \mid x) \quad (14)$$

$\phi(l)$ is a many-to-one map representing the set of CTC paths corresponding to l because multiple CTC paths can correspond to the same tag sequence. And $p(l \mid x)$ can be obtained by the forward backward algorithm.

The loss function of CTC is defined as the sum of the negative logarithmic probability marked by each training sample correctly, and the function is differentiable, so the CTC network can be trained by back-propagation algorithm:

$$\mathrm{L}_{CTC} = -\sum_{(x,l)} \ln p(z^l \mid x) = -\sum_{(x,l)} \mathrm{L}(x, z^l) \quad (15)$$

There are two main differences between CTC and the traditional label framework: 1) If the output is uncertain, additional blank label can be added to reduce the number of labels predicted by the network in one frame; 2) The training standard optimizes the logarithmic probability of the state sequence, rather than the input logarithmic likelihood.

3 Experimental Setup

The experiment uses the open source Mandarin voice corpus AISHELL-1 [7]. The corpus uses 3 recording devices for recording, and then selects high-fidelity microphone audio data and resamples to 16 kHz, 16 bits. In the corpus, the training set is 150 hours in total 120098 sentences, the test set is 5 hours and 7176 sentences, while the verification set is 10 hours in 14326 sentences which is used to cross-test the effect of training. The experiment used a character-level 5-gram language model with a size of 2.8GB, which was trained by Baidu's internally selected text collection [8].

The speech feature vector is generated by a Fourier-transform based filter-banks, which includes 40 coefficients distributed on a mel-scale and energy, along with their first and second temporal derivatives. The long-term characteristics of multi-frame series can effectively improve the performance of the model. Therefore, the long-term characteristics of a total of 11 frames consisting of 5 frames before and after the current frame are taken. All speech data are normalized by averaging over all training cases so that each coefficient or first derivative or second derivative all has zero mean and unit variance. According to the characteristics of Chinese character set, the network output is about 6,000 characters. Because of the instability of CTC training in the early stage, the probability of gradient anomaly in some long sentences is relatively high. The SortaGrad technique is used in the training. The initial values of the model parameters of the neural network are randomly selected from the uniform distribution range of [-0.1,0.1], and the initial learning rate is set to 5e-4.

In this paper, we use PReLU function. The Parametric Rectifier Linear Unit (PReLU) [9] is an extension of the ReLU in which the output of the model in the regions that input is negative is a linear function of the input with a slope of α. PReLU is formalized as:

$$\hat{H}_i = \begin{cases} H_i, & if\ H_i > 0 \\ \alpha H_i, & otherwise \end{cases} \quad (16)$$

The extra parameter α is usually initialized to 0.1 and can be trained using backpropagation.

In applications like speech recognition, we usually have access to the entire sequences. However, those sequences may have variable length. To solve this problem, we apply a sequence-wise normalization, where we compute the mean and variance of each feature along both the time and batch axis using:

$$\bar{x}_k = \frac{1}{n}\sum_{i=1}^{m}\sum_{t=1}^{T} x_{i,t,k} \tag{17}$$

$$\sigma^2_k = \frac{1}{n}\sum_{i=1}^{m}\sum_{t=1}^{T} (x_{i,t,k} - \bar{x}_k)^2 \tag{18}$$

BN introduces additional learnable parameters γ and β for scaling and moving data respectively:

$$BN(x_k) = \gamma_k \hat{x}_k + \beta_k \tag{19}$$

And the batch data processed by BN is used as a new input.

4 Experiment Result

The machine used for the experiment is an Intel Xeon E5-2640 v4 CPU, a Nvidia Tesla M40 GPU, and 128 GB of memory. Each set of experiments was run 3 times and the average was taken as the final experimental result.

4.1 Different LSTM Input

In previous experiments, CTC can reduce the length of input and output sequences at a lower frame rate, reduce the computational cost during decoding and provide an improvement in the delay. Specifically, three 10ms frames are directly stacked into the LSTM. After processing a stack frame once, 3 frames can be skipped. At the same time, the acoustic score evaluation during decoding occurs once every 30ms to speed up the process.

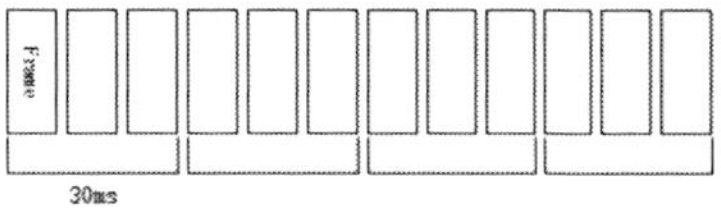

Figure 2. Stacking of input frames

The input of this model is first processed by CNN, each input is a single frame of 20ms, and then input to LSTM. In ASR, the signal characteristics in the low frequency region are very different from those in the high frequency region. Therefore, different spectral bands should use different filter weights so that it allows LFS to be used in convolution layers which weights span only a small local region of frequency. But in theory, given enough feature maps, FWS should be able to match the performance of LWS assuming the model is able learn which convolution feature maps to use and discard in the appropriate frequency bands.

So, this article applies FWS, uses one-dimensional and two-dimensional convolutions, respectively, where 1D is the convolution of the time axis and 2D is the convolution of the time and frequency axes. The stride of both is 1 and 2x1. The pooling layer is only added after the first convolution layer, only sampling the frequency, and its size is 1x3, while maximum sampling is used and does not overlap.

Results as shown in Table 1, there are three layers of LSTM with 1024 nodes in each layer and 3,386 nodes in the full connection layer. The recognition rate is word error rate (WER).

Table 1. Performance of different LSTM inputs

Input	Filters	Filter Size	WER	
			No BN	BN
Stack Frames	-	-	28.12 ± 0.02	25.59 ± 0.09
2 Layers-1D	512,512	5,5	24.53 ± 0.02	22.33 ± 0.07
2 Layers-2D	128,128	41x11 21x11	22.71 ± 0.11	20.68 ± 0.05

From Table 2, it can be seen that the two-dimensional convolution performance is better than one-dimensional convolution, and the recognition rate is improved by about 5.5% on average, and the final recognition result is improved. In low-resource corpus, limited data makes easier to under-fit and tends to favour common phoneme states with large probabilities, whereas two-dimensional convolutions can provide more efficient regularization. And the BN transformation reduces internal covariate shift by insulating a given layer from potentially uninteresting changes in the mean and variance of the layer's input which improves the final performance.

4.2 Pitch Features for Tonal Languages

In particular, the pitch features have been found to be beneficial for tonal languages (e.g., Mandarin, Cantonese and Vietnamese) [10]. On our Mandarin setup, we incorporate the pitch features into CTC model training. The pitch features are extracted using the method described in [10]. On each frame, appending the 3-dimensional pitch to the 40-dimensional filterbank features gives us a 43-dimensional feature vector. The experimental results in Table 2 show that the addition of tone features improves the performance of the system.

Table 2. Performance of adding pitch features

Model	Feature	WER
2 Layers-2D	filterbank	22.33 ± 0.05
	filterbank+pitch	21.66 ± 0.06

4.3 Decode

After the network trained by CTC, it needs an algorithm to decode it. The initial greedy search method which used above does not need to add any language information, selecting the maximum probability output of each time node to search for the best path.

In the process of CTC training, the acoustic model itself has the properties of the language model, and the decoding effect is good without adding the external language information, but the external language model trained with large text corpus is necessary to obtain the best results. In the decoding process, the beam search is used to find the optimal transcription y that formula 20 can obtain maximum.

$$Q(y)=\log(p_{ctc}(y \mid x))+\alpha \log(p_{lm}(y))+\beta word\_count(y) \quad (19)$$

The weight α controls the relative contributions of the language model and the CTC network. The weight β encourages more words in the transcription. These parameters are tuned on a development set.

Table 3. Decoding Performance for 2 Layers-2D

Method	LM	WER
Greedy Search	-	21.66± 0.06
Beam Search	5-gram LM	15.99± 0.08

From Table 3, we can see that the performance of the original greedy search method is obviously worse than beam search, and the recognition rate of beam search method with LM is increased by about 30%. The reason is that greedy search doesn't add any language information to create transcriptions. When in a slightly more complex situation, it is prone to disorder of words.

In formula 21, the optimization of the two hyper-parameters has a significant effect on the final decoding effect, so a two-dimensional grid search is performed to adjust the optimal value based on the validation set.

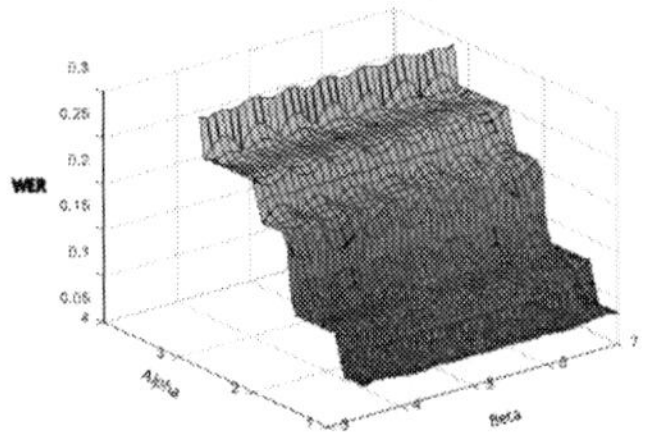

Figure 3. The 2D grid search for hyper-parameter

The grid search diagram for α and β is shown in figure 3. For LM, the global optimal value of WER is α=1.2、β=3.5, then decode the test set again.

Table 4. Decoding Performance for 2 Layers-2D

State	α , β	WER
Initial	2.6, 5.0	15.99± 0.08
Tune	1.2, 3.5	14.57± 0.03

The optimized α , β achieves better performance than the initial value on the test set, and the recognition rate is increased by 8.9%. It shows that the optimized super parameters can distribute weights more effectively and rationally, which is helpful to improve the performance of ASR.

Table 5. The Performance of Models

Model	WER
GMM-HMMs	21.73 ± 0.11
DNN-HMMs	14.82 ± 0.03
CNN-HMMs	14.16 ± 0.06

4 Conclusion

In this paper, an end-to-end speech recognition model for Mandarin is built. The input feature of the model uses convolution neural network instead of direct input and it has been modified and adjusted. We use the batch normalization and carry out two-dimensional convolution of input features in time domain and frequency domain. Finally, the effect of super-parameter optimization on the recognition effect is verified, improving the robustness of the model and achieves good results better than the mainstream hybrid ASR (Table 5).

References

1. Hinton, G., Deng, L., Yu, D., Dahl, G. E., ... & Kingsbury, B, *IEEE Signal Processing Magazine*, **29(6)**, 82-97, (2012)
2. Abdel-Hamid, O., Deng, L., & Yu, D, *Interspeech*, **Vol. 2013**, pp. 1173-5, (2013)
3. Hochreiter S, Schmidhuber J, *Neural computation*, **9(8)**, 1735-1780, (1997)
4. Graves A, Fernández S, Gomez F, *Proceedings of the 23rd international conference on Machine learning*, ACM, 369-376, (2006)
5. Graves A, Jaitly N, *International Conference on Machine Learning*, 1764-1772, (2014)
6. Sak H, Senior A, Beaufays F, *Computer Science*, 338-342, (2014)
7. Bu H, Du J, Na X, arXiv preprint arXiv: **1709.05522**, (2017)
8. Amodei D, Ananthanarayanan S, Anubhai R, *International Conference on Machine Learning*, 173-182, (2016)
9. He K, Zhang X, Ren S, *Proceedings of the IEEE international conference on computer vision,* 1026-1034, (2015)
10. Ghahremani P, BabaAli B, Povey D, *Acoustics, Speech and Signal Processing (ICASSP)*, 2494-2498, (2014)

Semi-empirical Theory of Distributions of Users of Social Online Networks by the Number of Contacts

Ibragim Suleimenov[1] and Akhat Bakirov[1]

[1]*Almaty University of Power Engineering and Telecommunications, Almaty, Republic of Kazakhstan*

Abstract. Based on the empirical analysis of contacts in the social network VKontakte, quantitative patterns are established that describe the distribution of users of social networks by the number of contacts (the number of friends). The parameter entering into the differential equation, whose solution is the proposed distribution, makes sense, similar to the Dunbar number and approximately coincides with it in magnitude. The fact that parameters characterizing a given distribution, depending on the size of the communication space in different cities, has been shown. However, the basic parameter that characterizes the fractal dimension of the communication space remains constant for all cities studied.

1 Introduction

The study of the communication space, formed by the rapid development of telecommunication technologies, is of considerable interest in several aspects.

Thus, a significant part of the campaigns for the promotion of goods on the market are currently conducted in social online networks [1-5], and such networks are already being used not as an auxiliary tool for campaigning, but as the main one [6].

Social online networks linking a large number of people are an ideal platform for carrying out various information impacts of a mass nature, including in the context of information wars [7-9].

At the same time, modern telecommunication technologies make it possible to receive online education, which creates a number of advantages (availability of university education courses around the world, saving time for moving, convenience for people with high demands) [10-12]. Mass open online courses (MOOK) are beginning to gain popularity [13-15].

The consistent use of social online networks for various purposes requires their adequate theoretical description at a quantitative level.

In this paper we analyze the distribution of users of social online networks by the number of friends (mutual contacts) and a model describing this distribution is proposed.

2 Experimental

Data used in the work were collected on the basis of direct collection of information on the number of friends of users of the social online network "VKontakte".

A sample of users residing in the following cities was randomly created: Almaty, Moscow, Kiev, Pavlodar, Novosibirsk. Counted the number of friends each user has. Based on these data, the dependencies shown in Fig. 1 - 5 (points). The abscissa is the number n, which are at the center of the intervals on which the axis is divided. The scale of the partition was chosen unevenly. The ordinate is the ratio of the total number of users, whose number of friends fit into this interval, to the extent of this interval. This construction approximately corresponds to the density of the distribution of users by the number of friends.

It can be seen that the curves obtained for various cities in which the Runet is actively exploited have a similar character.

3 Discussion

3.1 Theoretical model

The distribution of the number of users by the number of friends in the general case can be found on the basis of the following system of equations

$$\frac{dn_j}{dt} = n_{j-1}\sum_{i=0} w_{j-1,i}n_i - n_j\sum_{i=0} w_{j,i}n_i - \frac{n_j}{\tau}, \quad (1)$$

where n_j is the number of users whose number of friends is j, $w_{j,i}$ is the frequency of the formation of a new connection between persons having friends j and i, τ is the lifetime of the network user (the actual time of use of the resource).

Assuming that the functions under consideration with a change in the number j vary slowly, we can proceed to the continuous form of writing equation (1). Replacing the discrete numbers j and i by continuous variables x and y, we can write

$$\frac{d}{dt}n(x) = n(x-\Delta x)\int w(x-\Delta x, y)n(y)dy - \\ - n(x)\int w(x,y)n(y)dy - \frac{n(x)}{\tau}. \quad (2)$$

Applying the expansion in a Taylor series up to quadratic terms, we have

$$\frac{d}{dt}\frac{n(x)}{\Delta x} = -n(x)\int \frac{\partial}{\partial x}w(x,y)n(y)dy - \\ -\frac{dn}{dx}\int w(x,y)n(y)dy - \frac{1}{\tau}\frac{n(x)}{\Delta x}. \quad (3)$$

From this we obtain that for the stationary case

$$\frac{df}{dx} = f(x)K(x), \quad (4)$$

where

$$K(x) = \frac{\left[\int \frac{\partial}{\partial x}w(x,y)f(y)dy - \frac{1}{\tau}\right]}{\int w(x,y)f(y)dy}. \quad (5)$$

It is essential that the functions entering the multiplier for the function f (x) in the expression (4) are determined by the entire profile of the distribution under consideration.

Expression (4) allows us to interpret the establishment of an equilibrium distribution of users according to the number of friends through a stream directed towards their increase. This, in particular, allows us to use heuristic considerations when finding the function K (x).

These considerations are as follows. The literature currently uses the Dunbar number [16-19], which represents the maximum number of permanent social contacts (links) that one person can support. Initially, this number was obtained when studying the influence of the size and development of the neocortex on the size of the monkeys flock [16-17], but now it is increasingly used to describe the communication space [20-21].

It can be assumed that the Dunbar number describes the potential causing the flow in the positive direction along the x axis in the model under consideration. If the number of friends of this user is less than the Dunbar number, then one can expect that the derivative in formula (4) will have a positive sign (and vice versa).

The simplest entry satisfying this requirement has the form

$$\frac{df}{dx} = (a-x)f^{\alpha}, \quad (6)$$

where the factor α takes into account the fractal nature of the communication space.

This character of the communication space follows from obvious considerations. The formation of communication between two persons is facilitated if they have a common friend, i.e. there is a certain chain connecting them. Inclusion of such chains leads to a power-law dependence in formula (6).

Equation (6) is of first order and is easily integrable, we have

$$-(1-\alpha)f^{1-\alpha} = (a-x)^2 + C, \quad (7)$$

where C is the integration constant.

Further

$$f = \frac{A}{((a-x)^2+C)^{\beta}} \quad (8)$$

where the normalization factor A is introduced into the notation,

$$\beta = \frac{1}{\alpha - 1};\ \alpha = 1 + \frac{1}{\beta} \quad (9)$$

3.2 Comparison with experimental data

A comparison of the experimental results with the theoretical calculations based on the proposed model is presented in Fig. 1-5.

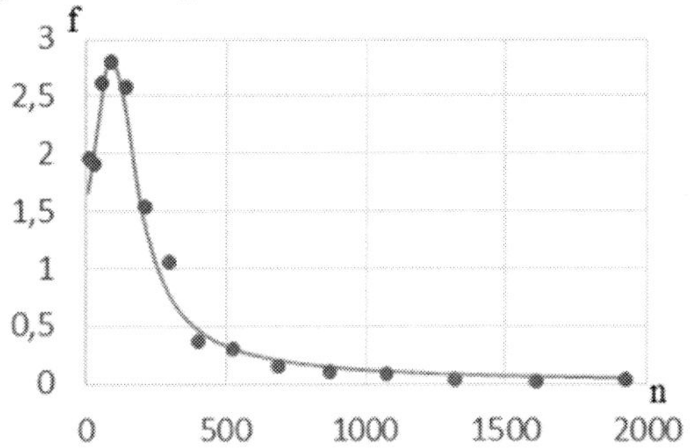

Figure 1. Comparison of experimental data with the theoretical calculation for Almaty, the sample is 3000 users.

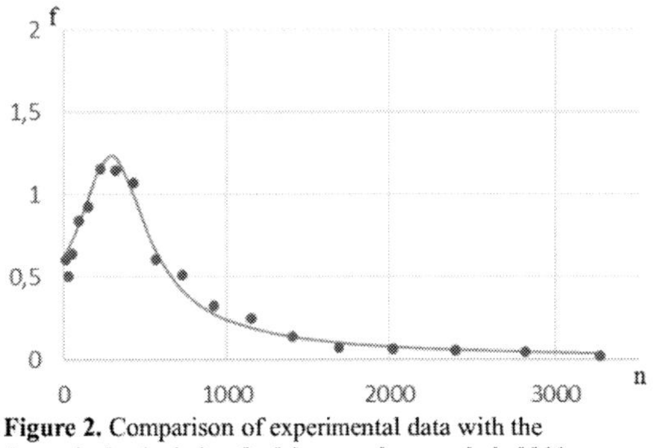

Figure 2. Comparison of experimental data with the theoretical calculation for Moscow, the sample is 3000 users.

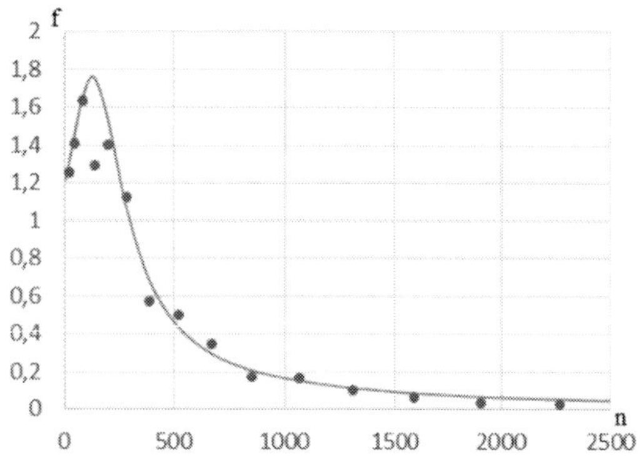

Figure 3. Comparison of experimental data with the theoretical calculation for Kiev, the sample is 3000 users.

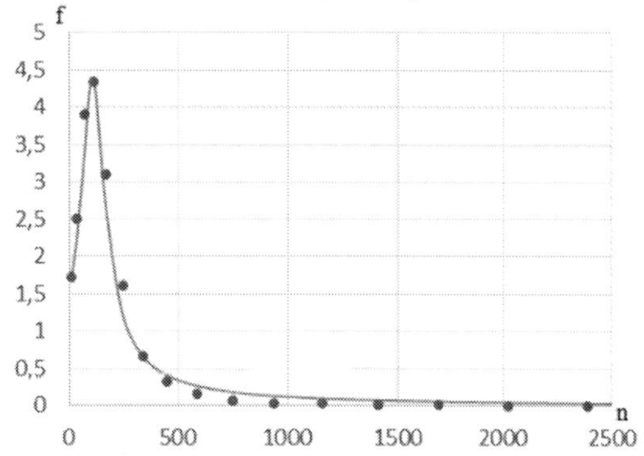

Figure 4. Comparison of experimental data with the theoretical calculation for Pavlodar, the sample is 3000 users.

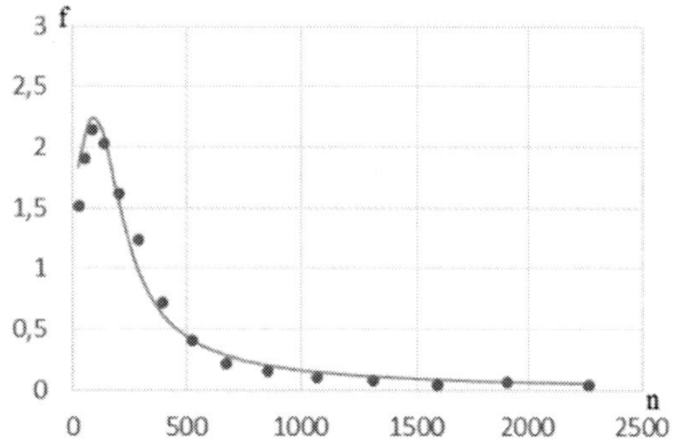

Figure 5. Comparison of experimental data with the theoretical calculation for Novosibirsk, the sample is 3000 users.

The theoretical curves (solid lines) in these figures were obtained on the basis of the model described above. It can be seen that there is good agreement between the experimental and theoretical data.

The control parameters that define the theoretical curve were determined on the basis of experimental data using the method of least squares. The numerical values obtained, which are used to construct the theoretical curves presented in Fig. 1 - 5, are summarized in Table 1.

It can be seen that analogues of the Dunbar number, obtained on the basis of the analysis of the presented dependences, fit within the range (100-230) indicated in the literature data [16-19]. It is also seen that the analog of the Dunbar number for Moscow proves to be much higher than for Almaty and Novosibirsk, which could be expected from general considerations.

The most significant result is that the fractal dimension of the communication space is the same for all three cities, with a value of 2/3 with high accuracy. It can be assumed that this value reflects some fundamental features of the formation of the communication space.

Table 1. Empirically determined constants of formula (8) for different cities.

City	A	C	a	β
Almaty	820	6000	95	0,654
Moscow	1290	44000	290	0,657
Kiev	1100	20000	130	0,656
Pavlodar	800	3000	110	0.648
Novosibirsk	1120	14000	95	0,651

4 Conclusions

Thus, on the basis of heuristic considerations and literary data on the nature of the Dunbar number, it is possible to propose a simple semi-empirical model that adequately describes the experimental distributions of users of social online networks by the number of friends. This model is based on the assumption of the fractal nature of the communication space. This assumption seems justified also because the fractal dimension obtained for various cities in which Runet is actively used remains with a high accuracy constant. The next step in the study is to expand the sample to more cities, and also to test for another segment of the Internet (not Runet)

References

1. W. Chen, C. Wang, Y. Wang. Proceedings of the 16th ACM SIGKDD international conference on Knowledge discovery and data mining. *Scalable influence maximization for prevalent viral marketing in large-scale social networks*. p. 1029-1038 (2010).
2. S. Doyle. Journal of Database Marketing & Customer Strategy Management. *The role of social networks in marketing*. **1**, p. 60-64 (2007).
3. M. Stelzner. Social media examiner. *2014 Social Media Marketing Industry Report*. p. 1-52 (2014).
4. D. Chaffey. Smart Insights: Social Media Marketing. *Global social media research summary 2016* (2016).
5. F. Eggers et al. Journal of small business management. *Technologies that support marketing and market development in SMEs—evidence from social networks*. **2**, p. 270-302 (2017).

6. E. Turban et al. Springer. *Electronic Commerce 2018: A Managerial and Social Networks Perspective* (2017).
7. E. Hunter, P. Pernik International Centre for Defence and Security. *The challenges of hybrid warfare* (2015).
8. J. Klausen Studies in Conflict & Terrorism. *Tweeting the Jihad: Social media networks of Western foreign fighters in Syria and Iraq*. **1**, p. 1-22 (2015).
9. Y. R. Kamalipour, M. Friedrichsen. Introduction: Transformation in Journalism and News Media. *Digital Transformation in a Global World*. p. 1-4 (2017).
10. L. de-Marcos et al. Computers & Education. *An empirical study comparing gamification and social networking on e-learning*. **75**, p. 82-91 (2014).
11. L. de-Marcos et al. Computers in Human Behavior. *Social network analysis of a gamified e-learning course: Small-world phenomenon and network metrics as predictors of academic performance*. **60**, p. 312-321 (2016).
12. P. De Meo et al. Information Sciences. *Combining trust and skills evaluation to form e-Learning classes in online social networks*. **405**, p. 107-122 (2017).
13. C. Bremer, D. Univ. Bibliothek Frankfurt am Main. *Weiß How to analyze participation in a (C) MOOC?p*. 992-1002 (2013).
14. M. Khalil, H. Brunner, M. Ebner. International Journal of Emerging Technologies in Learning (iJET). *Evaluation grid for xMOOCs*. **4**, p. 40-45 (2015).
15. J. F. Colas, P. B. Sloep, M. Garreta-Domingo. The International Review of Research in Open and Distributed Learning. *The effect of multilingual facilitation on active participation in MOOCs*. **4** (2016).
16. R. I. M. Dunbar. Behavioral and brain sciences. *Coevolution of neocortical size, group size and language in humans*. **4**, p. 681-694 (1993).
17. R. Dunbar. Harvard University Press. *Grooming, gossip, and the evolution of language* (1998).
18. R. I. M. Dunbar. Annals of human biology. *The social brain hypothesis and its implications for social evolution*. **5**, p. 562-572 (2009).
19. S. Shultz, R. I. M. Dunbar. I Know What You Are Thinking: Brain Imaging and Mental Privacy. *The social brain hypothesis: an evolutionary perspective on the neurobiology of social behaviour*. p. 13-28 (2012).
20. W. He. Networked Public. *Social Media: Tools and Space for Networked Public Communication*. p. 123-167. (Springer, Berlin, Heidelberg, 2017).
21. S. Banerjee. Interdisciplinary Description of Complex Systems: INDECS. *A biologically inspired model of distributed online communication supporting efficient search and diffusion of innovation*. **1**, p. 10-22 (2016).

Neural Networks and the Philosophy of Dialectical Positivism

Ibragim Suleimenov[1], Aliya Massalimova[2], Akhat Bakirov[1] and Oleg Gabrielyan[3]

[1]*Almaty University of Power Engineering and Telecommunications, Almaty, Republic of Kazakhstan*
[2]*Al-Farabi Kazakh National University, Almaty, Republic of Kazakhstan*
[3]*V.I. Vernadsky Crimean Federal University, Simferopol, Russian Federation*

Abstract. The fact that the theory of neural networks permits the completeness of the concept of global evolutionism has been shown. This concept in the current philosophical literature is seen, inter alia, as an effective platform for interdisciplinary cooperation, the need for which is becoming more acute, which is reflected in the anniversary report of the Club of Rome in the form of the thesis on the "New Enlightenment". The theory of neural networks allows us to give a consistent interpretation of the category of "complex", in accordance with which a system of arbitrary nature is treated as "complex" if it is possible to indicate a complementary analog of a neural network. With this interpretation, the evolution of systems of an arbitrary nature can indeed be described in a uniform way. In particular, the philosophical law of transition from quantity to quality can be reduced to a description in terms of information theory (through the description of the evolution of a neural network complementary to a complex system). The main result of the work is a new interpretation of the dialectical philosophy categorical apparatus on the basis of the theory of neural networks.

1 Introduction

At the present time, the issue of overcoming interdisciplinary barriers, which significantly reduces the effectiveness of scientific research worldwide, is becoming increasingly acute. Moreover, the trends associated with a significant increase in the number of scientific disciplines, an increase in the volume of scientific information, etc. lead to quite definite crisis phenomena that caused the authors of the jubilee report of the Club of Rome [1] to raise the question of the "New Enlightenment". In [1] it is recognized that the transition "from the consideration of reality as a whole to its division into many small fragments", once placed in the basis of the philosophy of the science of the New Time, no longer meets the current needs of civilization. An urgent necessity is the formation of a new philosophy of science as the foundation of the knowledge economy.

The Roman Club, once integrated into the global agenda of environmental discourse, retains its position as the elite of the world expert community. Its theses reflected and reflect a very specific social order, so there is reason to talk about the realized need for a radical change in the role of philosophy in society. The task of a new synthesis of scientific knowledge cannot be solved without an adequate philosophical understanding, respectively, the Renaissance of philosophical knowledge is inevitable.

At the same time, the problem under consideration cannot be solved by means of philosophy alone. The corresponding base is created within the natural sciences. It was on this basis that the concept of global evolutionism was constructed [2,3], based on the assumption that the laws of the evolution of systems of different nature are common.

In this paper it is shown that the theory of neural networks allows us to give a new interpretation of the concept of global evolutionism. It is de facto one of the prerequisites for the further development of the philosophy of dialectical positivism, one of the basic concepts of which is the concept of metalanguage (including the metalanguage of science).

2 Neural networks and the concept of global evolutionism

Analogies between neural networks and systems that are realized in nature and society, as shown in [we] are sufficiently deep. In particular, they allow us to justify the existence of the communication shell of a complex system as some new quality that appears in a complex system of any nature and that controls its behavior, including evolution. In section 2.1 we consider a concrete example of the appearance of such a quality, the nature of which does not reduce to the properties of the individual elements constituting the system.

2.1 Neural networks in nature and society

In works [4-6] it was shown that analogues of neural networks are found in nature and society much more often than it may seem at first glance.

In particular, in [5,6] considered an example of voting in a certain Council. Each member of the Council can be placed in accordance with the analog of the neuron. The analogy is legitimate, since each member of the Council actually converts the input signal (the information it receives, for example, the dissertation report) into a binary variable (if the "abstain" situation is excluded); "0" = "For"; "1" = "Against". However, in practice, each of the members of the Council takes into account, in one way or another, the opinion of the other colleagues. For example, a situation is common where the voice "Against" is given if the dissertant is a pupil of a competitor or an opponent. (It's easy to imagine a reverse situation.)

Schematically, the influence of council members on each other can be represented through a system of feedbacks, Fig.1.

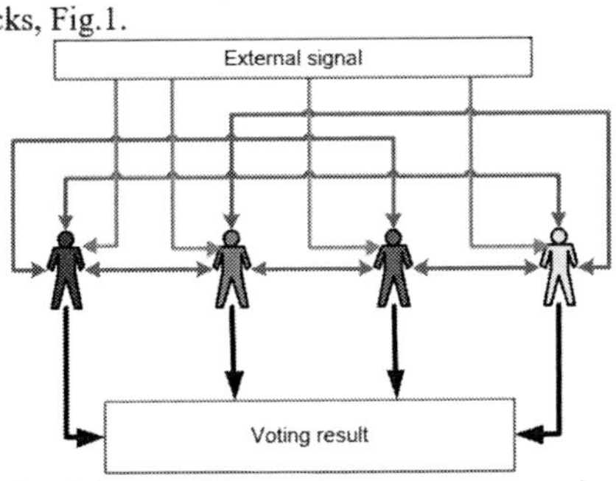

Figure 1. Scheme of interactions between members of the voting council [5]

The circuit in Fig. 1 is topologically equivalent to the Hopfield neuroprocessor and is described via a matrix of weight coefficients (Fig. 2).

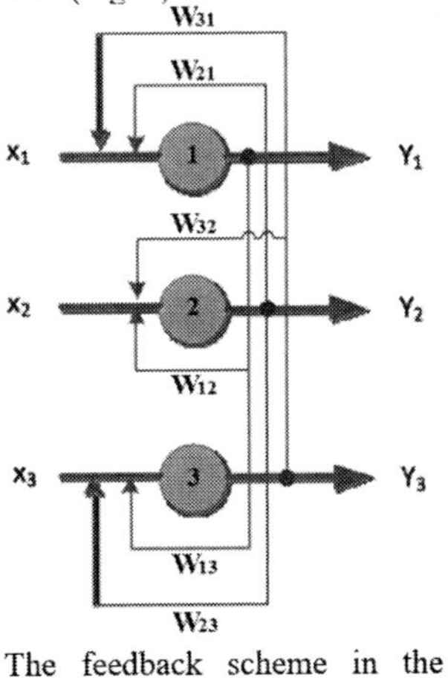

Figure 2. The feedback scheme in the Hopfield neuroprocessor

This example is important because it clearly shows that the decision is actually made not by individual members of the Council, or even by their aggregate, but by the neural network formed by them (if the number of members is large enough, and the connections between them are distributed). In other words, this example shows that there are processes of processing information that occur at a higher level than that associated with an individual element of the system. A new quality (communication shell) is emerging in the system, irreducible to the behavior of the individual constituent elements.

2.2 Neural networks and the category of complex

Consideration of objects similar to the voting Council and the generalization of the results obtained allowed us to work out the following interpretation of the category of the complex [4,7]. The system should be considered as complex when there is a complementary analog of the neural network. In other words, any complex system has its own communication shell, the processes in which influence its behavior, including evolution.

Namely, consideration of the communication shells of complex systems makes it possible to propose a mechanism for their evolution, an alternative to what goes back to the theory of the origin of C. Darwin's species. In accordance with this point of view, the evolution of complex systems proceeds in two stages. At the first stage, the structure of connections between the elements of the system takes place, which is interpreted as the evolution of its communication shell, which has relatively independent behavior. At this stage, the properties of the individual elements do not change.

At the next stage, the system as a whole is converted into an analog filter, which creates favorable conditions for the appearance of elements whose properties are more in line with the new state of the system. This approach allows to overcome many contradictions inherent in evolutionary theories based on the Darwinian point of view [4].

The theory of neural networks is important from this point of view to justify the concept of global evolutionism. However, with respect to the purposes of this work, another aspect is also important: the behavior of a complex system is controlled by a different quality (called the communication shell), which forms links between the elements and which is only indirectly related to the specific properties of the individual elements of the system.

For languages, including for the language of science, such a communication shell is the extra text structures discussed in the next section.

3 Languages and extra text structures from the point of view of the theory of neural networks

The history of the development of positivism consists of several stages. The latest of these stages is somehow connected with postpositivism, the development of which, in turn, is largely associated with the ideas of K. Popper.

According to Popper, science in general cannot deal with truth, since research activity is reduced to the promotion of hypotheses about the world, assumptions and conjectures about it. Judgments of this kind in [8] were interpreted as a kind of scientific decadence, the very emergence of which in the 20 th century serves as a graphic illustration of the growing negative trends in the development of world science, which eventually raised the question of the "New Enlightenment" [1].

Dialectical positivism does not share the views of postpositivists and continues the tradition of neo-positivism, overcoming the inherent contradictions on the basis of the dialectical method.

3.1. Modern reading of the ideas of neopositivism

Representatives of logical positivism (empiricism), in particular M. Shlik, R. Carnap, G. Reichenbach, focused their attention on the linguistic (with some simplification) problems of science [8].

There were quite certain grounds for this. In particular, the development of mathematics clearly showed that the language need not necessarily be natural, there are also artificial languages, in particular, the language of formal logic.

Further, everything that is done in science, in the final analysis, is expressed in certain linguistic forms, more precisely in the form of sentences of some language. Moreover, logic provides a convincing example of the existence of a language, the rules of operating which give reliable results. The question is, what then should the theory of knowledge as a whole differ from the discipline studying languages (both artificial and natural) as a means of constructing judgments? Indeed, if any theory is a certain system of judgments logically (or otherwise) related to each other (at least, ideally), therefore, at a minimum, there must be rules according to which the theory is constructed. But, if the theory is a means of cognition, then to an even greater extent the means of cognition are those tools with which it is used - that is, the means of the language, united by well-defined rules of operation. The developed logic of science (in the broad sense of the term) was called, in the opinion of neopositivists, to replace not only traditional philosophical ontology, but also traditional epistemology, the theory of knowledge.

We emphasize that the construction of neopositivists is currently acquiring an unexpected sound, in particular, in connection with the problems of developing a fuzzy logic apparatus [9]. (Fuzzy logic makes an attempt to link natural language and artificial, in particular, it actively uses such a concept as "linguistic variables.")

This circumstance - the existence of fuzzy logic, as well as other logics that do not coincide with the classical one, - highlights an important component of the reasoning of neopositivists. They, in fact, never claimed that classical logic, formalized by the works of mathematicians, is the only possible language of science. The apparatus of mathematical logic was really an ideal in their eyes, according to the canons of which the language of science should be built, but they clearly understood that the language of science is not limited to statements formalized by the simplest means.

Rather, neo-positivism formulated a program aimed at overcoming speculative philosophy, on the principles of developing a language of science broader than classical logic, but built on the same basic ideas. They saw one of the most important tasks of philosophy in creating a theory of scientific knowledge of a very specific type.

It is in this sense that dialectical positivism continues the tradition of neo-positivism, although its provisions are far from being reduced to solving problems that are somehow connected with the theory of knowledge. However, it is the analysis of the problems of the "language of science" that allows us to demonstrate the possibilities of the method of dialectical positivism, as well as to reveal the most important philosophical aspects of the theory of neural networks.

3.2 The problem of undecidable concepts from the point of view of dialectical positivism

The question of the basic concepts of logic has never been finally solved, which is demonstrated, in particular, by numerous attempts to construct a propositional calculus that operates modality categories, which include, in particular, the works of K. Lewis [10] and his followers [11,12].

From the most general positions, logic can be considered as the result of the formalization of one or another method of reasoning, which can be constructed according to various schemes. From this point of view, it is quite logical to develop alternatives to classical logic, which will increasingly approach the kind of reasoning that real human thinking uses.

Studies in the field of modal logic are actively pursued in our days [4, 12]. Common to them is the expansion of the classical logic of propositions and the classical logic of predicates, when the language of logic is replenished by modal operators of necessity and possibility acting on language sentences.

In a simplified form, the state of affairs can be expressed as follows. Classical logic is the method of reasoning that is most simple for formalization, the way that has been largely transferred to computers, only because formalization tools have already been well developed and mastered.

Evidence of this is the scheme of constructing a conceptual apparatus, applied in any of the scientific disciplines. This scheme is based on the construction of a system of definitions that make it possible to reduce the system of concepts used to basic concepts.

However, the development of the theory of neural networks [13] led to the appearance of the opposition "sequential calculations - parallel computations". Logic, one of the main tools of which is the construction of new true statements as consequences of already existing ones, is obviously associated with sequential computations. The theory of neural networks, the formation of which was originally associated with attempts to reveal the mechanisms of the functioning of human consciousness, unambiguously interprets the human brain as a system that realizes parallel computations.

From this point of view, in particular, it is of interest to analyze the existing approaches to the construction of axiomatics in any scientific discipline. The classical scheme of constructing axiomatics implies the existence of basic concepts through which all the others are expressed (more precisely, there is a certain hierarchy of concepts that allows one group of concepts to be consistently reduced to another). It is permissible to put such a scheme in line with the logic on which the sequential calculations are based.

However, this approach has a very definite drawback. Namely, it does not allow us to define (in the classical sense) the basic concepts. An example in this respect are the difficulties associated with the interpretation of the concept of "information" [4]. The definitions of information used in modern textbooks (information is the content of messages and the like) are not constructive; in fact, they are a tautology [14].

Objective dialectics finds a way out of this situation, defining the basic concepts through opposition. This is how pairs of categories of objective dialectics "contents-form", "quantity-quality", etc. are defined. In [4] arguments were presented proving that the concept of information should also be defined through opposition, treating this category as a pair to the category of matter.

3.3 The conceptual apparatus as a neural network

By appealing to the opposition "sequential computations - parallel computations", one can argue that the approach used by objective dialectics admits a definite generalization. Indeed, the opposition can be regarded as a special case of the connection between concepts.

Accordingly, it can be argued that it is permissible to define the basic concepts through a system of connections between them, and the opposition is considered no more than a particular case of the relationship established between the concepts. In other words, the concepts forming the basis form a network structure, which makes it legitimate to compare with parallel calculations performed by neural networks. Slightly exaggerating, we can say that a very specific neural network - the human brain - also uses a network conceptual structure to reflect the world around it, in which concepts are defined through connections with each other.

We emphasize that such an approach, in a certain sense, contradicts the foundations of classical logic. Namely, the use of a system of concepts that is self-enclosed can be interpreted as a "vicious logical circle". This, however, cannot be considered an obstacle for the following reasons. First, as emphasized above, the question of the foundations of logic is by no means closed. Secondly, this approach is much closer to everyday thinking than classical formal logic.

Hence, we can formulate the following problem. There is a certain language formed by a certain set of concepts and relations between them. Obviously, not all concepts in this language are basic, in any case, this is exactly what corresponds to the general case of "language".

The question is whether there are formalized tools that ensure the allocation of a basic set of concepts that form a network structure that defines them through interrelationships. Part of the task of this kind is solved by modern mathematical linguistics. For example, the problem of determining the vocabulary minimum necessary to explain in a foreign language was repeatedly discussed; the question is raised about the development of various variants of the "algebra of concepts" or "the calculus of concepts" [15-17]. However, the methods of solving such problems are far from complete formalization.

At the same time, the question of creating an algebra of concepts has one more dimension, which makes it necessary to consider it in a philosophical way.

Namely, any system that has passed a certain critical threshold of complexity, for example, a neural network, generates a different quality that is irreducible to the properties of the aggregate of constituent elements, as was shown in the previous sections. The most significant example is human consciousness - a new quality that appears in a system of related nerve cells, only because they are able to exchange signals that are electrical in nature.

Consequently, there is reason to believe that the basic structure of a fairly complex language, i. language, providing the formation of meaningful judgments, should also form some other quality, not reducible to individual concepts or their collections. In this respect, it is pertinent to emphasize that, as applied to natural language, the question of extra-textual structures, about the meanings that cannot be reduced directly to the text, has been discussed for a long time [18]. As noted by MK Mamardashvili, speaking about the subject of philosophy, "to feel those living things that stand behind the text and because of which, in fact, it arises. These things usually die in the text, they look badly through it, but nevertheless they are. "

It is much less obvious that extra-textual structures are equally inherent in artificial languages, for example, programming languages. Simplifying, the extra-textual structure of the programming language is its internal logic, the author's intention, which allowed transforming the original idea into a specific algorithmic structure. This circumstance allows us to say that the systems of concepts that generate a new quality are constructed in the same way as algebraic structures are constructed.

Representations of extra-textual structures as a new quality that forms the system of words of a language (the system of its concepts) correspond with the notions of a new quality that the neural network generates as follows. In both cases, we are talking about the emergence of a new quality, which is a purely systemic effect. It can be said that something else that appears in complex systems through the exchange of information between elements (say, through the exchange of signals between neurons of the brain) cannot have other reflection, except similar in information properties. Specifically, here we are talking about that new quality, which gives rise to a connection between the concepts of a particular language, it is not important whether it is artificial or natural.

In accordance with the proposed interpretation of the category of complex (section 2.2), the language can also be considered as a "complex system," which also must have its own communication shell, formed by what can be called meta-concepts. Such representations allow to speak correctly about "extra text structures", i.e. about an object that was previously studied only by means of the humanities. Meta-concepts are inexpressible in the language itself, but it is their existence that determines its logic. Consequently, the construction of any calculus of concepts that to some extent approximates reasonings in natural language must take into account the existence of meta-concepts.

4 Conclusions

Thus, the philosophical interpretation of the theory of neural networks, as a minimum, makes it possible to raise the question of consistent interpretation of the character by "extra textual structures" understood in the spirit of Yu.M. Lotman and U. Eco. The out-text structure in this interpretation is nothing more than a new quality that generates an analogue of a neural network formed by concepts and not reducible to the aggregate of individual elements of the network (in this case, individual concepts and / or utterances of natural or artificial language).

It is in this vein that we should also interpret the importance of analogies with neural networks for the philosophy of dialectical positivism, as a direction that focuses on the use of metalanguage forms.

References

1. E.U. von Weizsäcker, A. Wijkman. *Come On!: Capitalism, Short-termism, Population and the Destruction of the Planet.* (2017)
2. N.Z. Alieva, Z.V. Zayats. Mediterranean Journal of Social Sciences. *Paradigm Foundations of Modern Science: Self-Organization, Evolution, Integrity.* **3**, p.97 (2015).
3. G.P. Menchikov, B.Z. Sharifullin. The Social Sciences. *Global Evolutionism and Heterarchical Thinking.* **6**, p. 1250-1254 (2015).
4. I.E. Suleimenov, I.T. Pak, G.A. Mun. *THEORY OF EVOLUTION OF COMPLEX SYSTEMS AND ITS APPLICATIONS (In russian).* (2018)
5. I. Suleimenov, O. Gabrielyan, G. Mun, S. Panchenko, T. Amirzhan & K. Suleimenova. Int. J. on Communications. *Voting Procedure and Neural Networks.* **3**, p. 16-20 (2014).
6. I.E. Suleimenov, G.A. Mun, P.E. Grigoriev, S.V. Panchenko, D.B. Shaltykova & K.I. Suleimenova. Scientific notes of the Tavrida National University. VI Vernadsky *SOME EVIDENCE OF THE END OF THE FIRST PHASE OF THE NOOSPHERIC REVOLUTION IN THE CONTEMPORARY SOCIETY (In russian).* **26(5)**, p. 87-97 (2013).
7. Ye.M. Yergozhin, I.E. Aryn, G.A. Suleimenov, N.M. Mun, O.A. Belenko, N.T. Gabrielyan, El.-S. Negim. Park. In Seoul: Hollym Corporation Publishers. *Nanotechnology versus the global crisis.* (2010).
8. I.E. Suleimenov, O.A. Gabrielyan, Z.Z. Sedlakova, G.A. Mun. *History and philosophy of science: a textbook (In russian).* (2018)
9. F. A. Lootsma. Springer Science & Business Media. *Fuzzy logic for planning and decision making.* **8** (2013).
10. C.I. Lewis. The Journal of Philosophy. *A Pragmatic Conception of the A Priori.* **7**, p. 169–177 (1923).
11. M. Sinclair. *The Actual and the Possible: Modality and Metaphysics in Modern Philosophy.* (2017)
12. B. Kment. *Modality, metaphysics, and method //The Palgrave Handbook of Philosophical Methods.* – p. 179-207 (2015).
13. C.R. Alavala. *Fuzzy logic and neural networks: basic concepts & application.* (2008)
14. L. Brillouin. *Science and information theory.* (2013)
15. W.V. Quine. The Journal of Symbolic Logic. *Toward a calculus of concepts.* **1**, p. 2-25 (1936).
16. U. Mönnich. Meaning, Use, and Interpretation of Language. *Toward a Calculus of Concepts as a Semantical Metalanguage.* p. 1-35 (1983).
17. Y. Wang. International Journal of Cognitive Informatics and Natural Intelligence (IJCINI). *On concept algebra: A denotational mathematical structure for knowledge and software modeling.* **2**, p. 1-19 (2008).
18. Yu.M. Lotman. *Text and extra-textual artistic structures.* (1994)

Approximation properties of a new generalized Bernstein-Kantorovich operators

Bo-yong Lian

Department of Mathematics, Yang-En University, Quanzhou 362014, P.R.China

Abstract. By means of construction of suitable functions and the method of Bojanic-Cheng, the author gives the rate of convergence of a new generalized Bernstein-Kantorovich operators for some absolutely continuous functions.

1 Introduction

For a function $f(x)$ defined on the closed interval $[0, 1]$, the expression

$$B_n(f, x) = \sum_{k=0}^{n} f(\frac{k}{n})p_{n,k}(x)$$

is called the Bernstein polynomial of order n, where $p_{n,k}(x) = \binom{n}{k}x^k(1-x)^{n-k}$. The polynomials $B_n(f, x)$ were introduced by S.Bernstein [1] in order to give an especially simple proof of the Weierstrass approximation theorem. Then many scholars have done a lot of relevant research work. Bernstein operators became popular for several reasons: (1) they are given explicitly and depend only on the values of a function for rational values of the variable. (2) they have various shape-preserving properties and provide the simplest means for the study of some problems. (3) they are easy to handle in computer algebra systems when the evaluation of f is difficult and time-consuming.

Lorentz [2] gave an exhaustive exposition of main facts about the Bernstein polynomials. He also discussed some of their applications in analysis.

Based on the arithmetic mean of the total variation sequence, the estimation of the convergence rate of the B_n for the bounded variation function was obtained by Cheng [3]. It is proved that the estimation is essentially the best possibility of the continuous point.

Bojanic [4] investigated the asymptotic behavior of B_n for some absolutely continuous functions whose derivatives are bounded variation functions.

King [5] defined a new type of Bernstein operators which preserve x^2. Quantitative estimates were compared with estimates of approximation by the class Bernstein polynomials B_n in [5].

In the field of approximation theory, the applications of q-calculus are new area in the last 30 years. The first q-analogue of the well-known Bernstein polynomials was introduced by Lupas in the year 1987. In 1997 Phillips considered another q-analogue of the classical Bernstein polynomials. Next, the q-operators have become the research object of many scholars[6].

Recently, Chen et al. [7] introduced a new family of generalized Bernstein operators based on a non-negative parameter $\alpha(0 \le \alpha \le 1)$ as follows:

$$T_{n,\alpha}(f, x) = \sum_{k=0}^{n} f(\frac{k}{n})p_{n,k}^{(\alpha)}(x), \quad x \in [0, 1], \qquad (1)$$

where

$$p_{1,0}^{(\alpha)}(x) = 1 - x, p_{1,1}^{(\alpha)}(x) = x,$$

$$p_{n,k}^{(\alpha)}(x) = [\tbinom{n-2}{k}(1-\alpha)x + \tbinom{n-2}{k-2}(1-\alpha)(1-x)$$
$$+\tbinom{n}{k}\alpha x(1-x)]x^{k-1}(1-x)^{n-k-1}$$

for $n \geq 2$ and $\binom{n}{k} = 0 (k > n)$. When $\alpha = 1$, the operators $T_{n,\alpha}$ reduces to the classical Bernstein operators $B_n(f, x)$.

In [7], the authors studied many approximaiton properties of $T_{n,\alpha}$ such as uniform convergence, rate of convergence in terms of modulus of continuity, voronovskaya-type asymptotic formula, and shape preserving properties.

To approximate Lebesgue integrable functions, Mohiuddine et al. [8] introduced the following integral modification of the operators $T_{n,\alpha}$:

$$K_{n,\alpha}(f, x) = (n+1)\sum_{k=0}^{n} p_{n,k}^{(\alpha)}(x)\int_{\frac{k}{n+1}}^{\frac{k+1}{n+1}} f(t)dt. \quad (2)$$

In [8], the uniform convergence of the operators and rate of convergence in local and global sense are studied.

The rate of approximation for some absolutely continuous functions whose derivatives are bounded variation functions is an intersting topic. This is mainly originated from Bojanic-cheng [4], then many scholars have done a lot of research in this field [9-16].

Base on this, this article studies the approximation of $K_{n,\alpha}$ for some absolutely continuous functions $f \in DBV[0, 1]$, where

$$DBV[0, 1] = \{f | f(x) = f(0) + \int_0^x h(t)dt\}$$

and $x \in [0, 1], h \in BV[0, 1]$.

Let

$$R_{n,\alpha}(x, t) = \sum_{k=0}^{n}(n+1)p_{n,k}^{(\alpha)}(x)\chi_k(t),$$

where $\chi_k(t)$ is the characteristic function of the interval $[\frac{k}{n+1}, \frac{k+1}{n+1}]$ with respect to $I = [0, 1]$.

By the Lebesgue-Stieltjes integral representations, we have

$$K_{n,\alpha}(f, x) = \int_0^1 f(t)R_{n,\alpha}(x, t)dt. \quad (3)$$

2 Some lemmas

The proof of our result are based on the following lemmas.

Lemma 2.1 ([8]) For $e_i = t^i, i = 0, 1, 2$, we have

$$K_{n,\alpha}(e_0, x) = 1, \quad K_{n,\alpha}(e_1, x) = \frac{nx}{n+1} + \frac{1}{2(n+1)},$$

$$K_{n,\alpha}(e_2, x) = \frac{n^2}{(n+1)^2}\left(x^2 + \frac{n+2(1-\alpha)}{n^2}x(1-x)\right)$$
$$+\frac{nx}{(n+1)^2} + \frac{1}{3(n+1)^2}.$$

Remark 2.1 By simple applications of Lemma 2.1, we get

$$K_{n,\alpha}(t-x, x) = \frac{1-2x}{2(n+1)},$$

$$\begin{aligned} & K_{n,\alpha}((t-x)^2, x) \\ = & \frac{n+2(1-\alpha)-1}{(n+1)^2}x(1-x) + \frac{1}{3(n+1)^2} \\ \triangleq & \eta_{n\alpha}^2(x). \end{aligned}$$

Lemma 2.2 When n sufficient large, we have

$$K_{n,\alpha}(|t-x|, x) \leq \eta_{n\alpha}(x). \quad (4)$$

Proof By Cauchy-Schwarz inequality, we have

$$\begin{aligned} & K_{n,\alpha}(|t-x|, x) \\ \leq & \sqrt{K_{n,\alpha}((t-x)^2, x)} \cdot \sqrt{K_{n,\alpha}(1, x)} \\ = & \eta_{n\alpha}(x). \end{aligned}$$

The last inequality is obtained by Lemma 2.1 and Remark 2.1.

Lemma 2.3

(i) For $0 \leq y < x < 1$, there holds

$$\widetilde{R}_{n,\alpha}(x, y) = \int_0^y R_{n,\alpha}(x, t)dt \leq \frac{\eta_{n\alpha}^2(x)}{(x-y)^2}. \quad (5)$$

(ii) For $0 < x < z \leq 1$, there holds

$$1 - \widetilde{R}_{n,\alpha}(x, z) = \int_z^1 R_{n,\alpha}(x, t)dt \leq \frac{\eta_{n\alpha}^2(x)}{(z-x)^2}. \quad (6)$$

Proof (i) By (3) and Remark 2.1, we get

$$\begin{aligned}
&\widetilde{R}_{n,\alpha}(x,y)\\
&= \int_0^y R_{n,\alpha}(x,t)dt\\
&\le \int_0^y \left(\frac{x-t}{x-y}\right)^2 R_{n,\alpha}(x,t)dt\\
&\le \frac{1}{(x-y)^2}\int_0^1 (t-x)^2 R_{n,\alpha}(x,t)dt\\
&= \frac{1}{(x-y)^2} K_{n,\alpha}((t-x)^2,x)\\
&= \frac{\eta^2_{n\alpha}(x)}{(x-y)^2}.
\end{aligned}$$

(ii) Using a similar method, we have

$$\begin{aligned}
&1-\widetilde{R}_{n,\alpha}(x,z)\\
&= \int_z^1 R_{n,\alpha}(x,t)dt\\
&\le \int_z^1 \left(\frac{x-t}{z-x}\right)^2 R_{n,\alpha}(x,t)dt\\
&\le \frac{1}{(z-x)^2}\int_0^1 (t-x)^2 R_{n,\alpha}(x,t)dt\\
&= \frac{1}{(x-y)^2} K_{n,\alpha}((t-x)^2,x)\\
&= \frac{\eta^2_{n\alpha}(x)}{(z-x)^2}.
\end{aligned}$$

3 Main results

Theorem *Let $f \in DBV[0,1]$. If $h(x+)$ and $h(x-)$ exist at a fixed point $x \in (0,1)$, then we have*

$$\left|K_{n,\alpha}(f,x) - f(x) - \frac{(1-2x)[h(x+)+h(x-)]}{4(n+1)}\right|$$

$$\le \left|\frac{h(x+)-h(x-)}{2}\right|\eta_{n\alpha}(x)$$

$$+ \quad \frac{2\eta^2_{n\alpha}(x)}{x(1-x)}\sum_{k=1}^{[\sqrt{n}]}\bigvee_{x-\frac{x}{k}}^{x+\frac{1-x}{k}}(\varphi_x) + \frac{1}{\sqrt{n}}\bigvee_{x-\frac{x}{\sqrt{n}}}^{x+\frac{1-x}{\sqrt{n}}}(\varphi_x),$$

where

$$\varphi_x(t) = \begin{cases} h(t)-h(x+), & x < t \le 1;\\ 0, & t = x;\\ h(t)-h(x-), & 0 \le t < x.\end{cases}$$

Proof Let f satisfy the conditions of Theorem, by using Bojanic-Cheng's method [4], we have

$$f(t) - f(x) = \int_x^t h(u)du, \qquad (7)$$

where

$$h(u) = \frac{h(x+)+h(x-)}{2} + \frac{h(x+)-h(x-)}{2}sign(u-x)$$

$$+\varphi_x(u) + \delta_x(u)\left[h(x) - \frac{h(x+)+h(x-)}{2}\right] \qquad (8)$$

and

$$\delta_x(u) = \begin{cases} 1, & u = x;\\ 0, & u \ne x.\end{cases}$$

$$sign(x) = \begin{cases} 1, & x > 0;\\ 0, & x = 0;\\ -1, & x < 0.\end{cases}$$

From (7), (8), and noting $\int_x^t sign(u-x)du = |t-x|$, $\int_x^t \delta_x(u)du = 0$, we find that

$$\begin{aligned}
& K_{n,\alpha}(f,x) - f(x)\\
= \;& K_{n,\alpha}(f(t)-f(x),x)\\
= \;& K_{n,\alpha}(\int_x^t h(u)du, x)\\
= \;& \frac{h(x+)+h(x-)}{2}K_{n,\alpha}(t-x,x)\\
+ \;& \frac{h(x+)-h(x-)}{2}K_{n,\alpha}(|t-x|,x)\\
+ \;& K_{n,\alpha}(\int_x^t \varphi_x(u)du, x).
\end{aligned}$$

By Remark 2.1 and Lemma 2.2, we have

$$\left|K_{n,\alpha}(f,x) - f(x) - \frac{(1-2x)(h(x+)+h(x-))}{4(n+1)}\right|$$

$$\le \left|\frac{h(x+)-h(x-)}{2}\right|\eta_{n\alpha}(x)$$

$$+\left|K_{n,\alpha}(\int_x^t \varphi_x(u)du, x)\right|. \qquad (9)$$

To complete the proof, we must estimate the term $K_{n,\alpha}(\int_x^t \varphi_x(u)du, x)$.

From (3), the term $K_{n,\alpha}(\int_x^t \varphi_x(u)du, x)$ can be stated as

$$\begin{aligned} K_{n,\alpha}\left(\int_x^t \varphi_x(u)du, x\right) &= \int_0^1 (\int_x^t \varphi_x(u)du) R_{n,\alpha}(x,t)dt \\ &= \int_0^1 (\int_x^t \varphi_x(u)du) d_t\widetilde{R}_{n,\alpha}(x,t) \\ &= \int_0^x (\int_x^t \varphi_x(u)du) d_t\widetilde{R}_{n,\alpha}(x,t) \\ &+ \int_x^1 (\int_x^t \varphi_x(u)du) d_t\widetilde{R}_{n,\alpha}(x,t). \end{aligned}$$

Let

$$\Delta_{1n} = \int_0^x (\int_x^t \varphi_x(u)du) d_t\widetilde{R}_{n,\alpha}(x,t),$$

$$\Delta_{2n} = \int_x^1 (\int_x^t \varphi_x(u)du) d_t\widetilde{R}_{n,\alpha}(x,t).$$

Then we have

$$K_{n,\alpha}\left(\int_x^t \varphi_x(u)du, x\right) = \Delta_{1n} + \Delta_{2n}. \tag{10}$$

Using partial integration and noticing $\widetilde{R}_{n,\alpha}(x,0) = 0$, $\int_x^x \varphi_x(u)du = 0$, we get

$$\begin{aligned} \Delta_{1n} &= \widetilde{R}_{n,\alpha}(x,t) \int_x^t \varphi_x(u)du\Big|_0^x \\ &- \int_0^x \widetilde{R}_{n,\alpha}(x,t)\varphi_x(t)dt \\ &= -\int_0^x \widetilde{R}_{n,\alpha}(x,t)\varphi_x(t)dt \\ &= -(\int_0^{x-\frac{x}{\sqrt{n}}} + \int_{x-\frac{x}{\sqrt{n}}}^x) \widetilde{R}_{n,\alpha}(x,t)\varphi_x(t)dt. \end{aligned}$$

Thus, it follows that

$$\begin{aligned} |\Delta_{1n}| &\le \int_0^{x-\frac{x}{\sqrt{n}}} \widetilde{R}_{n,\alpha}(x,t) \bigvee_t^x (\varphi_x)dt \\ &+ \int_{x-\frac{x}{\sqrt{n}}}^x \widetilde{R}_{n,\alpha}(x,t) \bigvee_t^x (\varphi_x)dt \end{aligned}$$

From Lemma 2.3 (i) and $0 \le \widetilde{R}_{n,\alpha}(x,t) \le 1$, we get

$$|\Delta_{1n}| \le \eta_{n\alpha}^2(x) \int_0^{x-\frac{x}{\sqrt{n}}} \frac{\bigvee_t^x(\varphi_x)}{(x-t)^2}dt + \frac{x}{\sqrt{n}} \bigvee_{x-\frac{x}{\sqrt{n}}}^x (\varphi_x). \tag{11}$$

Putting $t = x - \frac{x}{u}$ for the integral of (11), we get

$$\begin{aligned} \int_0^{x-\frac{x}{\sqrt{n}}} \frac{\bigvee_t^x(\varphi_x)}{(x-t)^2}dt &= \frac{1}{x}\int_1^{\sqrt{n}} \bigvee_{x-\frac{x}{u}}^x (\varphi_x)du \\ &\le \frac{2}{x}\sum_{k=1}^{[\sqrt{n}]} \bigvee_{x-\frac{x}{k}}^x (\varphi_x). \end{aligned}$$

From (11) and the above inequality, it follows that

$$|\Delta_{1n}| \le \frac{2\eta_{n\alpha}^2(x)}{x}\sum_{k=1}^{[\sqrt{n}]} \bigvee_{x-\frac{x}{k}}^x (\varphi_x) + \frac{x}{\sqrt{n}} \bigvee_{x-\frac{x}{\sqrt{n}}}^x (\varphi_x). \tag{12}$$

From Lemma 2.3 (ii), using the same method, we also get

$$|\Delta_{2n}| \le \frac{2\eta_{n\alpha}^2(x)}{1-x}\sum_{k=1}^{[\sqrt{n}]} \bigvee_x^{x+\frac{1-x}{k}} (\varphi_x) + \frac{1-x}{\sqrt{n}} \bigvee_x^{x+\frac{1-x}{\sqrt{n}}} (\varphi_x). \tag{13}$$

Theorem now follows from (9),(12) and (13). This completes the proof.

Acknowledgment

This work is supported by the Program for Outstanding Youth Scientific Research Talents in Fujian Province University(2016).

References

[1] S N. Bernstein, Démonstration du théorème de Weierstrass fondeé sur le calcul des probabilités. Common. Soc. Math. Kharkhow. **13**(2), 1-2(1912-13)

[2] G G. Lorentz, Bernstein Polynomials. Univ. of Toronto Press(1953)

[3] F. Cheng, On the rate of convergence of Bernstein polynomials of function of bounded variation. J. Approx. Theory. **39**, 259-274(1983)

[4] R. Bojanic, F. Cheng, Rate of convergence of Bernstein polynomials for functions with derivatives of bounded variation. J. Math. Anal. Appl. **141**(1), 136-151(1989)

[5] J P. King, Positive linear operators which preserve x^2. Acta. Math. Hungar. **99**, 203-208(2003)

[6] A. Aral, V. Gupta, R P. Agarwal, Applications of q-Calculus in Operator Theory. Springer New York(2013)

[7] X Y. Chen, J Q. Tan, Z. Liu, J. Xie, Approximation of functions by a new family of generalized Bernstein operators. J. Math. Anal. Appl. **450**, 244-261(2017)

[8] S A. Mohiuddine, T. Acar, A. Alotaibi, Construction of a new family of Bernstein-Kantorovich operators. Math. Meth. Appl. Sci. **40**(18), 7749-7759(2017)

[9] B Y. Lian, Rate of approximation of bounded variation functions by the Bézier variant of Chlodowsky operators. J.Math.Inequal. **7**(4), 647-657(2013)

[10] T. Acar, A. Kajla, Blending type approximation by Bézier-Summation-Integral type operators. Commun. Fac. Sci. Univ. Ank. Ser. A1 Math. Stat. **67**(2), 195-208(10)

[11] T. Neer, A M. Acu, P N. Agrawal, Bézier variant of genuine-Durrmeyer type operators based on Polya distribution. Carpathian J. Math. **33**(1), 73-86(2016)

[12] P N. Agrawal, N. Ispir, A. Kajla, Approximation properties of Lupas-Kantorovich operators based on polya distribution. Rendiconti del Circolo Matematico di Palermo Series 2. **65**(2), 185-208(2016)

[13] V. Gupta, D. Soybaş, Convergence of integral operator based on different distributions. Filomat. **30**(8), 2277-2287(2016)

[14] V. Gupta, A M. Acu, D F. Sofonea, Approximation of Baskakov type Polya-Durrmeyer operators. Appl. Math. Comput. **294**:318-331(2017)

[15] Q B. Cai, The Bézier variant of Kantorovich type λ-Bernstein operators. J. Inequal. Appl.(2018) 2018: 90. https://doi.org/10.1186/s13660-018-1688-9

[16] Q B. Cai, B Y. Lian, G. Zhou. J. Inequal. Appl.(2018) 2018: 61. https://doi.org/10.1186/s13660-018-1653-7

High-resolution Distributed ISAR Imaging based on Sparse Representation

Yuanyuan Li, Yaowen Fu and Wenpeng Zhang

College of Electronic Science, National University of Defense Technology, Changsha, China

Abstract. Distributed ISAR technique has the potential to increase the cross-range resolution by exploiting multi-channel echoes from distributed virtual equivalent sensors. In the existing imaging approaches, the echoes acquired from different sensors are rearranged into an equivalent single-channel ISAR signal. Then, the missing data between the observation angles of any two adjacent sensors is restored by interpolation. However, the interpolation method can be very inaccurate when the gap is large or the signal-to-noise (SNR) of echoes is low. In this paper, we discuss sparse representation of distributed ISAR echoes since the scattering field of the target is usually composed of only a limited number of strong scattering centres, representing strong spatial sparsity. Then, by using sparse algorithm (Orthogonal Matching Pursuit algorithm, OMP), the positions and amplitudes of the scattering points in every range bin can be recovered and the final ISAR image with high cross-range resolution can be obtained. Results show the effectiveness of the proposed method.

1 Introduction

The cross-range resolution of conventional inverse synthetic aperture radar (ISAR) is obtained by exploiting the relative rotation between the target and the radar [1,2]. To achieve a high cross-range resolution, a long observation time is needed, which will make the translation motion compensation complicated. In order to overcome this drawback, the antenna array can be used to obtain the image of a maneuvering target by one single snapshot [3]. However, this technique needs a large number of antenna elements, which is costly. To increase the cross-range resolution of ISAR image with limited antenna elements, the distributed ISAR technique has been considered [4-6]. This technique has the possibility of utilizing the data acquired by multiple radar sensors. Each sensor is characterized by either transmitting capability or receiving capability. The receiving sensors can receive and separate all the transmitting signals. Therefore, any transmitting sensor and any receiving sensor can form a virtual equivalent sensor that autonomously transmits and receives radar waveforms. The synthetic aperture formed by target motion is utilized to fill the gap between the adjacent virtual sensors. With appropriate radar formation, each scatter can be observed from much wider observation angles and thus can achieve a higher cross-range resolution.

A very significant step in distributed ISAR imaging is how to deal with the multi-channel echoes. In the existing approaches, the multi-channel echoes are rearranged according to the change of the observation angle [5,6]. Three basic cases are distinguished, (1) overlap in observation angle, (2) no overlap, with a small gap (sensors close), (3) no overlap, with a large gap (sensors far apart) [7]. In case (1), it needs time selection to reduce overlap. And in case (2), the gap filling interpolation technique can be utilized. While in case (3), the interpolation may not be suitable since the gap is large and interpolation may not be correct.

In this paper, we mainly focus on the case (3). While the interpolation technique may recover the missing data in case (2), its performance deteriorates when the gap of observation angle is large. Also, the interpolation technique is susceptible by noise. Therefore, a sparse imaging method is proposed to overcome these drawbacks. Meanwhile, the sparse imaging method has no sidelobe in the final image.

The rest of this paper is organized as follows. After presenting the signal model of distributed ISAR in Section 2, the sparse imaging method is introduced in Section 3. Then, simulation results are presented in Section 4 to validate the effectiveness of the proposed method. Finally, we conclude this paper in Section 5.

2 Distributed ISAR signal model

Consider a 2D coordinate system (O,X,Y) with the origin in the target's rotating centre. The target is modelled as a rigid body, consisting of Q scatters and rotating clockwise. The M transmitting sensors and N receiving sensors are placed in the XOY plane (see Figure.1).

Assume the M transmitting signals $s_m(t)$ are orthogonal, and each receiving sensor has the ability to

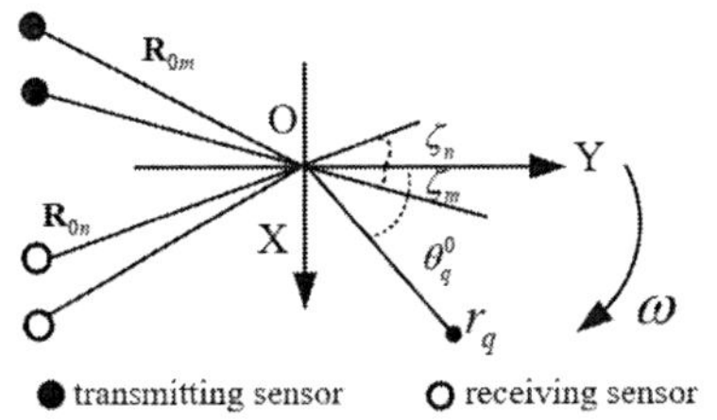

Figure.1. geometry of distributed ISAR imaging

receive and separate the signals from all the M transmitting sensors. After demodulation and filtering with matched filters in each receiving sensor, the received backscattered signal in the m-nth observation channel is denoted as

$$y_{mn}(\hat{t},t_p)=\sum_{q=1}^{Q}\sigma_{mn,q}p_m[\hat{t}-R_{mn,q}(t_p)/c]e^{-j\frac{2\pi}{\lambda}R_{mn,q}(t_p)} \quad (1)$$

where $\hat{t}$ is the fast time, $t_p = pT_p$ is the slow time, $T = PT_p$ is the accumulation time, c is the wave velocity, λ is the carrier wavelength, $\sigma_{mn,q}$ is the scattering coefficient of the qth scatter in the m-nth observation channel, and $R_{mn,q}(t_p)$ is the propagation distance from the sensor m to the scatter q and back to sensor n. $p_m(\hat{t})$ denotes the point spread function of $s_m(\hat{t})$; $p_m(t)\approx p(t),\forall m$, where $p(t)$ is the sinc function [6].

The position vector of the qth scatter can be written as $\mathbf{r}_q(t_p)=r_q[\sin(\theta_q^0+\omega t_p),\cos(\theta_q^0+\omega t_p)]$, where r_q is the vector length, θ_q^0 is the initial azimuth angle of the scatter, and ω is the constant rotation speed in clockwise. $\mathbf{R}_{0m}$ and $\mathbf{R}_{0n}$ are denoted as the position vectors of the mth transmitting sensor and the nth receiving sensor, respectively, which can be expressed as

$$\mathbf{R}_{0m}=R_{0m}[-\sin\xi_m,-\cos\xi_m] \quad (2)$$

$$\mathbf{R}_{0n}=R_{0n}[-\sin\xi_n,-\cos\xi_n] \quad (3)$$

where ζ_m and ζ_n are the azimuth angles measured clockwise. Under the far-field assumption, the propagation distance $R_{mn,q}(t_p)$ can be expressed approximately as

$$\begin{aligned}R_{mn,q}(t_p)&=\left\|\mathbf{R}_{0m}-\mathbf{r}_q(t_p)\right\|+\left\|\mathbf{R}_{0n}-\mathbf{r}_q(t_p)\right\|\\&=R_{0m}+R_{0n}-(\mathbf{R}_{0m}/R_{0m}+\mathbf{R}_{0n}/R_{0n})\mathbf{r}_q(t_p)\\&=2(R_{mn}+r_q\cos(\theta_q^0+\omega t_p-\alpha_{mn})\cos\beta_{mn})\end{aligned} \quad (4)$$

where the mean distance R_{mn}, the mean angle α_{mn}, and the half difference angle β_{mn} are equal to $R_{0m}+R_{0n}$, $(\zeta_m+\zeta_n)/2$, and $(\zeta_m-\zeta_n)/2$, respectively [5].

The transmitting/receiving couple (m,n) can be regarded as the ith virtual equivalent sensor, and the number of all equivalent sensors is I. By setting $\alpha_i=\alpha_{mn}$, $\beta_i=\beta_{mn}$ and neglecting the constant distance R_{mn} under the assumption that the translation motion has been compensated, $R_{mn,q}(t_p)$ can be written as

$$r_{iq}(t_p)=2r_q\cos(\theta_q^0+\omega t_p-\alpha_i)\cos\beta_i) \quad (5)$$

$r_{iq}(t_p)$ reflects the range compression position in range dimension and the phase signal in cross-range dimension. We will simplify $r_{iq}(t_p)$ in range dimension and cross-range dimension respectively.

In the range dimension, we assume the accumulation angle ωT is very short, so the range migration can be neglected with the approximations of $\sin(\omega t_p)\approx 0$ and $\cos(\omega t_p)\approx 1$ .Therefore, $r_{iq}(t_p)$ can be written as $(y_{q0}\cos\alpha_i+x_{q0}\sin\alpha_i)\cos\beta_i$ ((x_{q0},y_{q0}) is the initial coordinate of the qth scatter). Although we assume that the angle between the equivalent sensor is large, this is only relative to the accumulation angle. So, we assume that the α_i and β_i are not very large and have little influence on the range position after compression. So, it must satisfy $\left|y_{q0}(\cos\alpha_i-1)+x_{q0}\sin\alpha_i\right|\cos\beta_i<\rho_r$ (ρ_r is the range resolution). When $y_{q0}=0$, ζ_m and ζ_n must satisfy

$$\left|\sin\zeta_m+\sin\zeta_n\right|<\left|2\rho_r/x_{q0}\right| \quad (6)$$

For example, if $\rho_r=1$, $x_{q0}=20$, $\zeta_m=0$, the absolute value of ζ_n must less than 5.7^o , and α_i will less than 2.85^o . Actually, α_i cannot be very large in order to satisfy (6). Thus, for echoes of different equivalent sensors, each scatter is compressed into the same range bin. Therefore, from (5), $r_{iq}(t_p)$ can then be written as $2y_{q0}$, which simplifies the process in the range dimension.

Meanwhile, in the cross-range dimension, the observation angle can be regarded as $\omega t_p-\alpha_i$ .Since the value of t_p is $[-T/2,T/2]$, so $\omega t_p-\alpha_i$ is changed from $-T\omega/2-\alpha_i$ to $T\omega/2-\alpha_i$, which is shown in Figure.2. For the sake of simplicity, we assume that the relationship between the mean angle of each equivalent sensor is $\alpha_{i-1}>\alpha_i>\alpha_{i+1}$. In order to ensure there exists gap between the observation angles of two adjacent equivalent sensors, the α_i and the rotation angle ωT must satisfy $\alpha_{i-1}-\alpha_i>\omega T$.

Then, since α_i is very small, with the assumptions of $\sin\alpha_i\approx 0$, $\cos\alpha_i\approx 1$, $\sin(\omega t_p)\approx\omega t_p$ and $\cos(\omega t_p)\approx 1$, (5) can be expanded as

$$r_{iq}(t_p)=2[y_{q0}-x_{q0}\cos\alpha_i\omega t_p]\cos\beta_i \quad (7)$$

The first addend of (7) is constant and has no effect on the imaging result. Therefore, the received signal of the ith equivalent sensor can be expressed as

$$\begin{aligned}y_i(\hat{t},t_p)=&\sum_{q=1}^{Q}\sigma_{i,q}p[\hat{t}-2y_{q0}/c]\\&\cdot\exp(j\frac{4\pi}{\lambda}x_{q0}\omega t_p\cos\alpha_i\cos\beta_i))\end{aligned} \quad (8)$$

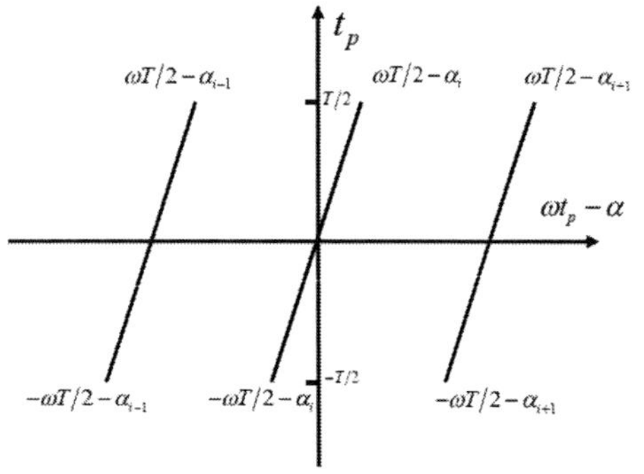

Now, the question is how to use the multiple echoes from all the equivalent sensors to increase the cross-range resolution. In existing approaches, the echoes of all equivalent sensors are rearranged according to the observation angle. After uniform interpolation to fill all the gaps, the standard ISAR focusing techniques can then be applied. However, this method cannot handle the case which the gap is large or the SNR of echoes is low. In this paper, we use sparse imaging method to get well focused result.

3 Sparse imaging method

If there is no overlap and no gap of the multiple echoes, the standard ISAR focusing method can be applied after rearranging these echoes. However, if the gap exists, it will be difficult to get the desired imaging result.

I equivalent sensors can obtain I echoes with $I-1$ blocks of missing data. The number of valid sampling points in cross-range dimension obtained by each equivalent sensor is P, the total number is $L_v = IP$. And the number of the sampling points to be recovered is L, $L > L_v$, that is, the imaging in this situation is an underdetermined problem. Modern compressive sensing theory has proved that if the original signal satisfies the sparse property under a certain observational base space that complies with the RIP criterion (Restricted Isometric Property), then the number of samples less than the original signal length can also recover the original sparse signal [8]. In general, the ISAR image appears as a superposition of scattering points in several different doppler frequency units in the cross-range. We consider dealing with the distributed ISAR signals with missing data in a sparsely optimized manner.

3.1. Sparse representation

We only consider the cross-range signal in one range bin, which is shown in Figure.3. According to (8), this signal for the ith equivalent sensor can be written as

$$s_i(t_p)=\sum_{q=1}^{Q}\sigma_{iq}\exp(j\frac{4\pi}{\lambda}x_{q0}\omega t_p\cos\alpha_i\cos\beta_i) \quad (9)$$

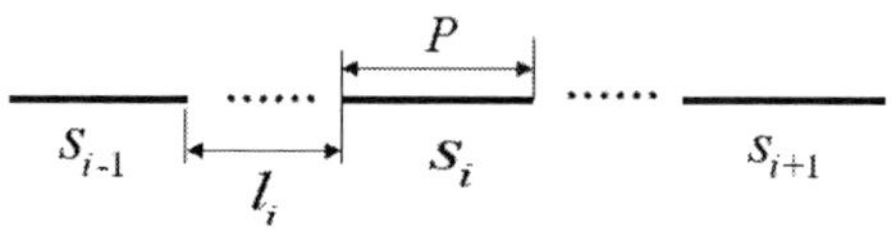

Figure.3. distributed ISAR sampling data

s_i has P sampling points. The whole valid sampling points for all echoes is $L_v = IP$. The number of missing data points between two adjacent sensors is

$$l_i=((\alpha_{i-1}-\alpha_i)/\omega-T_p)PRF \quad (10)$$

And the global signal length is

$$L=IP+\sum_{i=2}^{I}l_i \quad (11)$$

Combined with (9), the global azimuthal signal can further be expressed as

$$\mathbf{S}=\mathbf{\Phi}\boldsymbol{\sigma} \quad (12)$$

where $\mathbf{S}\in\mathbb{C}^{L_v\times 1}$, $\sigma\in\mathbb{C}^{L\times 1}$. $\mathbf{\Phi}$ is the overcomplete base space based on α_i, β_i, the target rotation speed ω and other parameters. The essence of $\mathbf{\Phi}$ is the Fourier transform matrix:

$$\mathbf{\Phi}=[(\mathbf{F}_1)^T\cdots(\mathbf{F}_i)^T\cdots(\mathbf{F}_I)^T]_{L_v\times L} \quad (13)$$

where the $\mathbf{F}_i$ is expressed as

$$\mathbf{F}_i=\begin{bmatrix}\exp(j2\pi f_1 t_1\cos\alpha_i\cos\beta_i) & \cdots & \exp(j2\pi f_1 t_P\cos\alpha_i\cos\beta_i)\\ \vdots & \vdots & \vdots\\ \exp(j2\pi f_L t_1\cos\alpha_i\cos\beta_i) & \cdots & \exp(j2\pi f_L t_P\cos\alpha_i\cos\beta_i)\end{bmatrix} \quad (14)$$

And f represents a uniformly varying doppler frequency which belongs to $[-PRF/2, PRF/2]$.

3.2. OMP method

Reconstruction of cross-range image can be regarded as the process of finding the solution to (12):

$$\min\|\bar{\sigma}\|_0 \quad s.t. \quad \mathbf{\Phi}\bar{\boldsymbol{\sigma}}=\mathbf{S} \quad (15)$$

Based on l_1 -norm, the (15) can be transformed to

$$\min\|\mathbf{S}-\mathbf{\Phi}\bar{\boldsymbol{\sigma}}\|_2+\lambda\|\bar{\boldsymbol{\sigma}}\|_1 \quad (16)$$

which is a nonlinear convex optimization problem.

Orthogonal Matching Pursuit Algorithm (OMP) is an improvement on matching pursuit algorithm [9,10]. OMP algorithm needs to orthogonalize the selected atoms by the Gram-Schmidt method, and then project the signals in the space formed by these orthogonal atoms to obtain the signal components on the selected atoms and the residual components. Then, use the same method to decompose the residual components. After V decomposition, the original signal is decomposed into a linear combination of V atoms. In each step of decomposition, the selected optimal atom satisfies certain conditions. Therefore, the residual component decreases rapidly with the decomposition process. In this way, the original signal can be represented by a small number of atoms and can be converged after a limited number of iterations.

Therefore, the imaging steps are shown in Fig.4 and explained as follows:

(1) rearrange echoes of all equivalent sensors after passing matched filters based on the observation angle
(2) coarse imaging to find where scatter points exist from all range bins
(3) select a range bin and extract this bin's cross-range signal as $\mathbf{S}$
(4) construct $\mathbf{\Phi}$ matrix
(5) obtain the result of $\hat{\mathbf{\sigma}}$ in this range bin with OMP method
(6) select another range bin's signal to solve

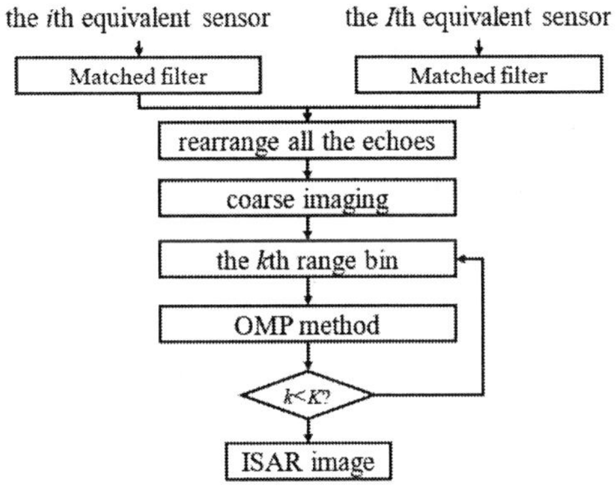

Figure.4 the whole processing chain

4 Simulation results

In this section, simulations are conducted to demonstrate the effectiveness of the proposed method. The distributed sensors are composed of two transmitting sensors and two receiving sensors ($M=2,\ N=2$). The azimuth angles of ζ_m and ζ_n are $(6^o,1^o)$ and $(-6^o,-1^o)$, respectively. Therefore, three equivalent sensors can be obtained with mean angles $\alpha_1=2.5^o$, $\alpha_2=0^o$, $\alpha_3=-2.5^o$ (actually four equivalent sensors are formed with two have the same mean angle 0°). The transmitting signal is a set of orthogonal signals with the same center frequency 10GHz and the same bandwidth 300MHz which can achieve a range resolution of 0.5m.

The target is composed of eight scattering points (see Figure.5) which are isotropic and independent with each other. Assume the target rotates uniformly with an angular speed of 0.02 rad/s. Since the mean angle α_i and angular speed ω are fixed, the gap between observation angles only depends on the accumulation time. 2s and 1.5s are considered separately. For the former case, the accumulation angle for single sensor is 2.3^o, and for the latter case, it is 1.72^o . In these two cases, the range compression echoes rearranged according to the observation angle are shown in Figure.6 (a) and Figure.6(b). So, the gap is larger when the accumulation time is shorter.

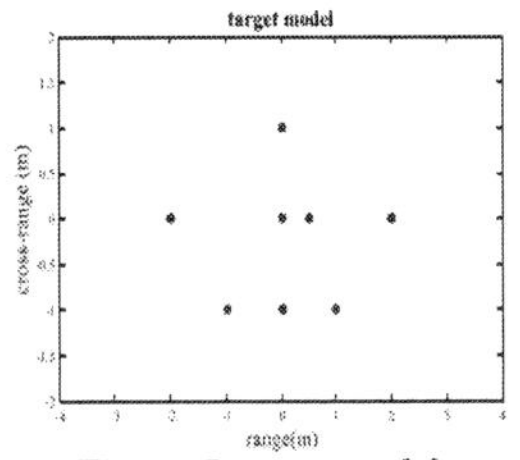

Figure.5. target model

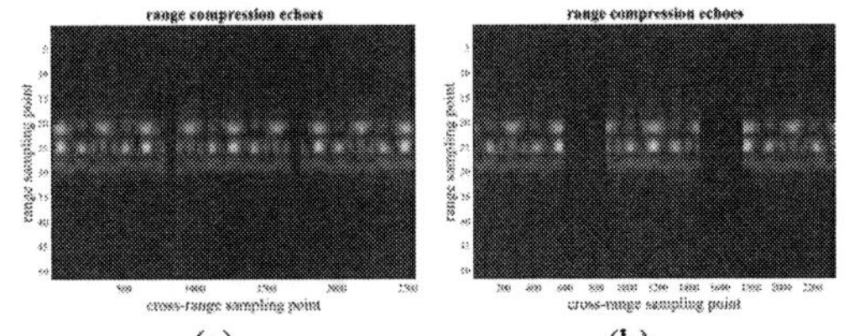

(a) (b)

Figure.6. range compression echoes

(a) $T=2s$ (b) $T=1.5s$

Firstly, we analyse the performance of interpolation technique in the two cases. Echoes in the 25th range bin which have four scatters are selected. In the first case, the number of valid sampling points is 3×801 , and 2×73 for the missing data. In the second case, the number of valid sampling points is 3×601 , and 2×273 for the missing data. The results of FFT on recovered echoes by interpolation are shown in Figure.7. When the accumulation time is 2s, the cross-range resolution is 0.375m for single sensor, so the two scatters $(0,0)$ and $(0.5,0)$ cannot be distinguished clearly. From Figure.7 (a), four scatters are all distinguished, which means the interpolation technique can recover the missing data well, and a higher resolution can be achieved by exploiting all echoes to increase the accumulation angle. While when the accumulation is 1.5s, that is, more missing data, the interpolation technique fails to fill the gap. So, the FFT cannot get good result.

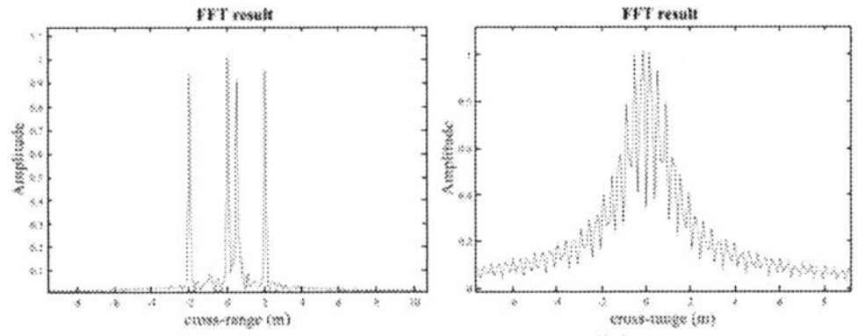

(a) (b)

Figure.7. FFT results after interpolation

(a) $T=2s$ (b) $T=1.5s$

Then, we observe the performance of OMP method. We use the echoes in the 25th range bin in the second case. The result with super-resolution feature is shown in Figure.8 (a). Compared with Figure.7 (b), OMP can get the estimated positions and amplitudes of all scatters. Figure.8 (b) is the final result of all selected range bin.

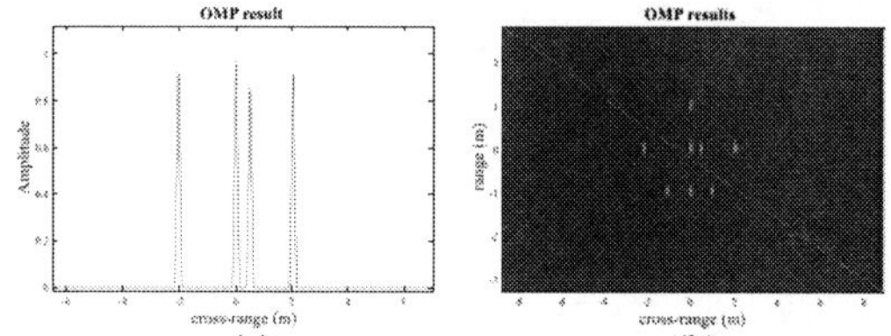

(a) (b)

Figure.8. OMP results ($T = 1.5s$) (a) the 25th range bin (b) all range bins

Lastly, we test the effect of noise. Echoes in the 25th range bin are used ($T = 2s$). White gaussian noise is added to make $SNR = 5dB$. FFT result is shown in Figure.9 (a), which is affected by the noise and cannot identify the positions of scatters, compared with Figure.7 (a). While from Figure.9 (b), the OMP method works well, the noise has little influence on the high-resolution result.

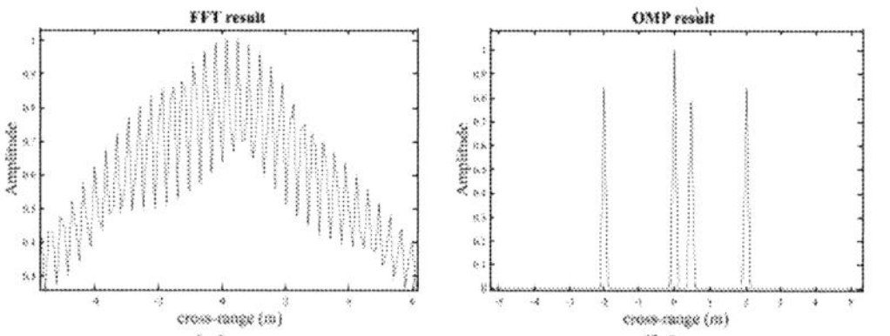

(a) (b)

Figure.9. comparison of two methods ($SNR = 5dB$) (a) FFT after interpolation (b) OMP

From the three simulations, we can obtain:

(1) the interpolation technique can fill the gap between adjacent observation angles when the gap is small. And by exploiting all the echoes to increase the accumulation angle, the cross-range resolution can be enhanced. However, it cannot get good results when the gap is large or the echoes' SNR is low.

(2) the OMP method performs well to get high-resolution result compared with the interpolation technique.

4 Conclusion

In this paper, a sparse method is proposed to get the cross-range image of distributed ISAR. Firstly, the signal model of distributed ISAR is introduced. The constraint condition is derived to make each scatter is compressed in the same range bin for echoes of different equivalent sensors. Secondly, the sparse representation of the cross-range signals shows that the sparse method can be used to get the estimated positions and amplitudes of scattering points based on Fourier transform matrix. The simulation results show that the OMP method can get high-resolution result without recovering the missing data. And the OMP method can work well when the gap is large or the echoes' SNR is low. However, when the migration through resolution cell exists between echoes of different equivalent sensors, the cross-range signal cannot be jointly processing. Therefore, the correction of range migration needs further research. Also, if the target has three-dimensional rotation, how to apply this sparse representation method also needs to be discussed.

References

1. J. L. Walker, "Range-Doppler imaging of rotating objects", IEEE Trans. Aerosp. Electron. Syst., **AES-16**, 23-52 (1980)
2. H. Q. Wu, D. Grenier, G. Y. Delisle, and D. G. Fang, "Translation motion compensation in ISAR image processing", IEEE Trans. Image Process., **4**, 1561-1571 (1995)
3. B. D. Steinberg, "Microwave imaging of aircraft", Proc. IEEE, **76**, 1578-1592, (1988)
4. P. Debora, M. Bucciarelli, and P. Lombardo, "Multi-platform ISAR for flying formation," in IEEE Radar Conf., 1-6, (2009)
5. P.Debora, M. Bucciarelli and P. Lombardo, "Multistatic and MIMO distributed ISAR for enhanced cross-range resolution of rotating targets," IEEE Trans. Geosci. Remote Sens., **48**, 3300-3317, (2010)
6. Y. T. Zhu, Y. Su, and W. X, Yu, "An ISAR imaging method based on MIMO techinique", IEEE Trans. Geosci. Remote Sens., **48**, 3290-3299, (2010)
7. P. V. Dorp, M. P. G. Otten and J. M. M. Verzeilberg, "Coherent multistatic ISAR imaging", Iet Int. Conf. on Radar. Syst. IET, 1-6, (2013)
8. D. L. Donoho, "Compressed sensing," IEEE Trans. Inf. Theory, **52**, 1289-1306, (2006).
9. J. A. Tropp, A. C. Gillbert, "Signal recovery from random measurements via Orthogonal matching pursuit", IEEE Trans. Inf. Theory, **53**, 4655-4666, (2007)
10. E. J. Candès, J. K. Romberg, T. Tao, "Stable signal recovery from incomplete and inaccurate measurements," Communications on Pure and Applied Mathematics, **59**, 1207-1223, (2006)
11. J. A. Tropp, A. C. Gillbert, "Signal recovery from random measurements via Orthogonal Matching Pursuit" IEEE Trans. Inf. Theory, **53**, (2007)

Research on the Trusted Enhancement Technology Based on Bare Metal Model Virtual Machine

Fei Wang, Yu Wang, Jie Qiang

Space Engineering University, Space Information Institute, Beijing, China

Abstract. This paper proposes a trusted enhancement model for bare-metal VMs, which can provide credible guidance and application credible enhancement in combination with the startup and booting characteristics of bare-metal VMs. The model is suitable for terminals with multi-security service requirements. It can be divided into several trusted virtual isolation terminals according to the actual needs of users. It has the characteristics of not changing the existing network topology, high security and rapid deployment.

1 Introduction

With the intensification of national informatization construction, the construction of various information systems is increasing day by day, and the access equipment and applications have become increasingly complex. Due to the different service types and security levels of the information system, users often need to operate multiple terminals at the same time, which makes it more and more difficult for a user to manage daily usage and operation and maintenance. Virtual machine technology can work in different virtual environments by separating different types of applications such as office, business processing, and information access to meet users' needs of safely handling different dense-level services in the same physical terminal environment. Despite all the advantages, virtual machine systems face the same security vulnerabilities and are vulnerable to code security threats [1].

Starting with the bare-metal virtual machine, this paper proposes a credible and enhanced technology based on the bare-metal virtual machine through the study of the start-up and operation mechanism of the bare-metal virtual machine, and establishes a credible and safe operating environment for the bare-metal virtual machine system , Which not only can provide users with technical support for the safe deployment and use of multi-level security business systems, but also is of great significance to the development of virtualization technology and the construction of important information systems.

2 Virtual machine technology

2.1 A Virtual machine classification and characteristics

The concept of virtual machines appeared in the 1960s. The virtual machine monitor (VMM) is an important part of the virtual machine system. It divides the entire computing platform into many isolated virtual machines [2]. The operating environment of each virtual machine is almost the same as that of a normal terminal operated by a user, and all the virtual machines have access to the hardware resources through the VMM. According to VMM, the virtual machine system can be classified into three types: host type, bare-metal type and hybrid type at different levels of the system [3], as shown in Fig. 1.

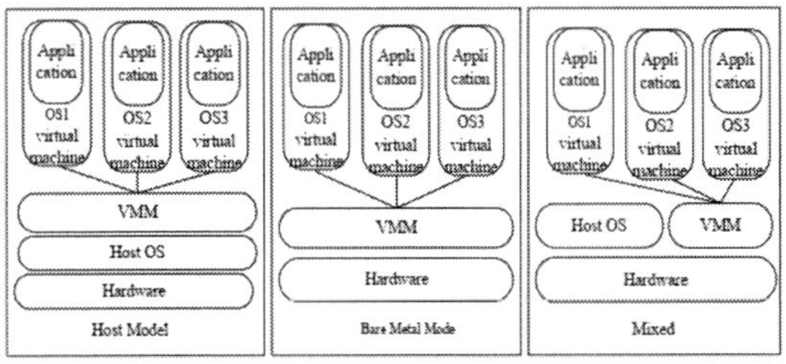

Figure 1. Virtual machine system classification.

(1) Host VMs VMM is located on the host operating system, so that the physical resources owned by the host operating system can be fully utilized. The VMM invokes the services provided by the host operating system to complete the virtualization work and multiple user operations The system runs on the same host operating system.

(2) VMM bare metal model is located in the hardware layer above, you can directly call the hardware interface, it controls and is responsible for managing all the hardware resources and the upper virtual machine environment. IO requests for each operating system are handled by the VMM.

(3) A hybrid virtual machine is a hybrid of the above two structures. The VMM is still created on the host operating system. It needs to modify the kernel of the host operating system to assume part of the VMM functions, and the VMM can directly access the hardware resources。

The three types of virtual machines each have their own characteristics. Specifically, the comparison between them is shown in Table 1.

Table 1. Virtual machine classification comparison.

	Host model	Bare model	Hybrid
Typical representative	VMware Workstation and so on	VMware EXS Server, WindRiver Hypervisor and more	Xen、Sun Logical Domain and more
Isolation	better	it is good	better
System performance	general	high	Higher
Trusted computing base size	Larger	Smaller	Larger
The main advantage	You can directly use the host operating system driver, IO device virtualization easier to achieve	VMM can control all hardware; System isolation, stability is better.	You can directly operate the hardware to improve the operating efficiency of the virtual machine
Mainly inadequate	The efficiency of resource access has declined	VMM itself is relatively large, need to include a large number of drivers	Need to make changes to the host operating system

From the above analysis, VMM in a bare-metal VM takes over the system boot and operation before the operating system starts up, thereby having better isolation and higher system performance. From the perspective of trusted computing, both host and hybrid virtual machines are built on the host operating system, and the trusted computing base is relatively large. Therefore, this subject chooses bare-metal virtual machine to carry out credible security enhancement research.

2.2 Bare-metal virtual machine startup process

Bare-metal VMs include components such as the VM monitor, management VMs, user VMs, and I / O VMs, and the startup process for the VMs is shown in FIG 2.

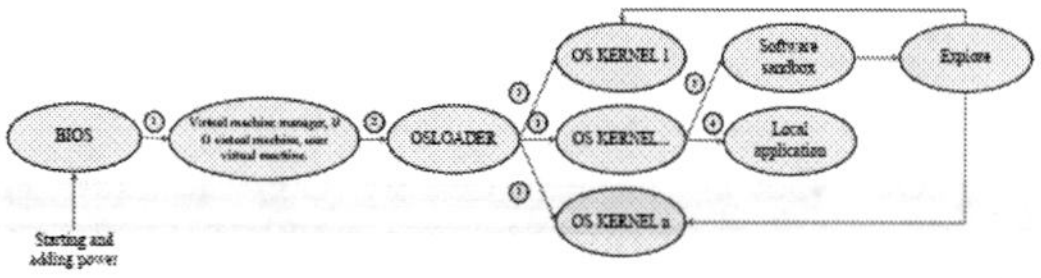

Figure 2. Virtual machine system classification.

Step 1: Power-on and power-on The BIOS loads the VMMs, user VMs, and I / O VMs respectively;

Step 2: VM Manager loads OSLOADER;

Step 3: According to the preset number of virtual machines, OSLOADER loads OS KERNEL for each user virtual machine respectively;

Step 4: Load the local application of the user VM according to the user selection or the latest usage, at this time, the user can use the terminal normally;

Step 5: If the user needs to switch to another virtual machine, the software box is loaded by the current virtual machine. After the user selects a virtual machine to be switched in the software sandbox, the software box loads Explore, Explore, and loads the selected virtual Machine applications and services, and the selected virtual machine interface to switch to the user view, the original virtual machine interface to switch to the background.

3 Trusted model based on bare-metal virtual machine

In the startup and booting phase of a bare-metal virtual machine, there are more entities involved in the system than the ordinary PC. In addition to the boot program and the operating system, the system also includes a virtual machine monitor, an IO virtual machine, and a management virtual machine. During the operational phase, a collection of applications that need to maintain multiple virtual machines introduce new vulnerabilities, both in terms of program count and complexity. On the one hand, the introduction of these special modules will reduce the credibility of the system; On the other hand, during system operation, attackers can more easily exploit vulnerabilities in a certain module to implant malicious programs such as viruses and Trojan horses and cause damage to the system. Bare metal virtual machine to achieve credible security enhancement, we must guide from the system to the actual operation of business software, fully guarantee the credibility of the system.

3.1 Virtual machine trust transfer model

Definition 1:

S Virtual machine boot process all entities collection, $s \in S$;

H The set of expected summary values for all entities in the virtual machine's boot process, $h \in H$;

R The integrity status of the entity. $R = \{trusted, untrusted\}$;

In the system before the first start, that the system is credible, calculated S The expected summary values for all entities in Generate Collection H , And save it at credible root.

Definition 2:

Summary value calculation function $hash: S \rightarrow H$, Generates a digest value for each entity calculation.

Integrity verification function $validate: S \times hash(S) \times H \rightarrow R$,for

$\exists s_i, s_j \in S$,by s_i an examination s_j The integrity of the meet expectations.

If $hash(s_j) = h_j$,then $validate(s_i, hash(s_j), h_j) = trusted$,otherwise $validate(s_i, hash(s_j), h_j) = untrusted$.

Definition 3:

for $\exists s_i, s_j \in S$,

if $validate(s_i, hash(s_j), h_j) = trusted$,then s_j Is credible, and vice versa s_j Is not credible. This is also the method used by the TCG in establishing the chain of trust to indicate whether it is trustworthy by measuring the integrity value of the data.

Definition 4:

symbol $\xrightarrow{-}$ Indicates the direct trust relationship between two entities. for $\exists s_i, s_j \in S$,

$s_i \xrightarrow{-} s_j$ Said s_i Direct trust s_j

symbol $\xrightarrow{\bar{S}}$ Represents the transfer of trust between two entities. for $\exists s_i, s_j \in S$, $s_i \xrightarrow{\bar{S}} s_j$ Said s_i Through the entity sequence $\bar{s}$ Indirect trust s_j

Definition 5:

function $parent(S) \to S$ Represents the parent entity that acquired the entity. Correct $\forall s_i, i \geq 2$,

$\exists s_j$ Make $s_j = parent(s_i)$.

Definition 6:

function $\Omega(S \times S) \to \bar{S}$ Represents a function that gets the sequence between the upper entity and the lower entity,

where the collection $\bar{S} \subseteq S$.

for $\exists s_i, s_j \in S, i > j$ If $\Omega(s_j, s_i) = \phi$, Then said s_i versus s_j There is no physical sequence between,

i.e. $s_j = parent(s_i)$ Otherwise $\Omega(s_j, s_i) = \bar{s} \neq \phi$,then said s_i versus s_j There is some entity sequence between $\bar{s}$

Theorem 1:

In the model, s_1 Is credible root. The virtual machine boot process is credible, if and only if it is satisfied $\forall s_i \in S, \exists s_j = parent(s_i)$,

$\therefore$ Virtual machine boot process in line with the basic idea of trust transfer, boot process is credible.

$validate(s_j, hash(s_i), h_i) = trusted$

prove:

$\because s_1$ Trusted root

$\therefore s_1$ Unconditionally credible

$\because \forall s_i \in S, i \geq 2$, Parent entity $parent(s_i)$ Perform an integrity check, $validate(hash(s_i), h_i) = trusted$

$\therefore$ By definition 1-3, set S Any entity is credible

$\because$ According to definitions 1-6, $\Omega(parent(s_i), s_i) = \phi$

$\therefore parent(s_i) \xrightarrow{-} s_i$

$\because \forall s_i \in S$,

Equation $s_1 = parent(\cdots parent(s_i))$ Set up, and $\Omega(s_1, s_i) = \bar{s}$

$\therefore \forall s_i \in S$, $s_1 \xrightarrow{-} s_i$ or $s_1 \xrightarrow{\bar{S}} s_i$

3.2 Based on the bare metal model of trusted enhanced architecture

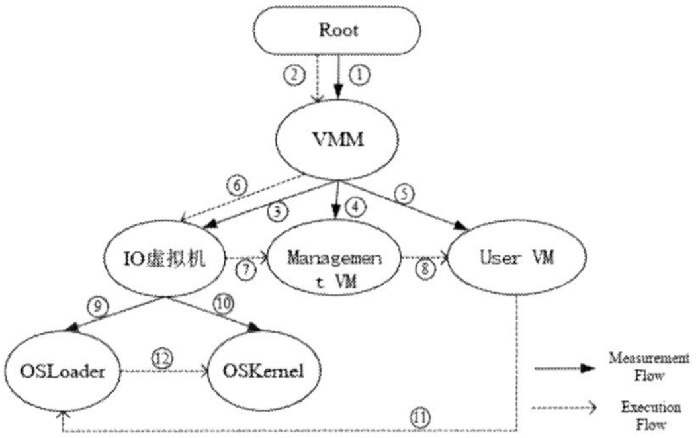

Figure 3. Trusted and enhanced architecture based on bare-metal VMs.

According to the trust transfer model of virtual machines and the startup loading order of bare-metal VMs, the trusted enhanced architecture of bare-metal VMs is shown in Fig. 3. Trusted root measures VMM integrity, verify that the VMM is trusted, and start execution

(1) Credible root Firstly, the trustworthiness of VMM is measured and loaded after verification.

(2) After the VMM is started, measure the IO virtual machines, manage the virtual machines, and then verify the integrity of the user VMs. After the VMs are verified, they are loaded and executed sequentially through the IO VMs, the management VMs, and the user VMs.

(3) After the IO virtual machine starts, the OSLoader and the OSKernel integrity are measured sequentially, and the OSLoader and the OSKernel are loaded and executed after the verification is successful.

(4) The user virtual machine loads the parent operating system and their respective incremental parts, and the boot process is completed.

4 Prototype system design and safety analysis

4.1 Prototype system design

Trusted prototype system based on bare-metal virtual machine is composed of seven functional modules such as bare-metal virtual machine, Ukey credible root, authentication, trust establishment in boot phase, trust establishment in operational phase, security audit and security policy, Figure 4 shows.

(1) Bare-metal virtual machine. According to user needs from the underlying hardware virtual multi-user virtual machine.

(2) Ukey credible root. Is a trusted root for a bare-metal virtual machine and is responsible for storing confidence metrics, identity certificates, providing cryptographic services for trusted metrics, and identity authentication.

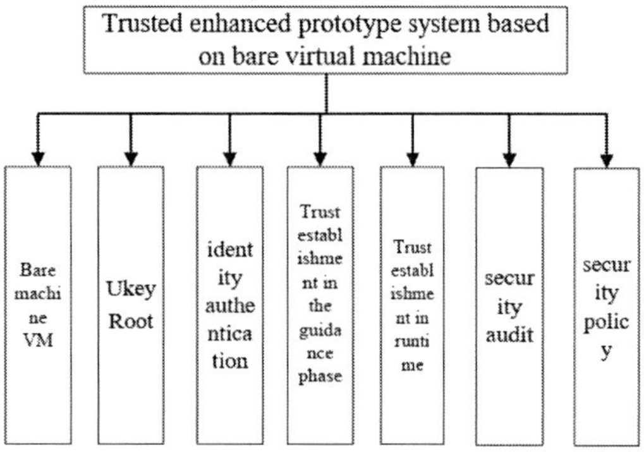

Figure 4. Prototype system functional structure.

(3) Identity authentication function. In the system start-up phase, the authentication function mainly completes the identification of the terminal user identity to prevent illegal users from accessing the system.

(4) Guide the establishment of trust. In this phase, the process of constructing the trust chain in the guiding phase is implemented, providing a credible operating environment for the virtual machine.

(5) Trust in the operational phase of the establishment. Mainly based on the white list of application layer executable program integrity verification.

(6) Security audit. Conduct security audits of user actions of trusted virtual machines.

(7) Security strategy. Maintaining and managing the bare metal trusted virtual mechanism security policy.

4.2 Security analysis

Trusted prototype system based on bare-metal virtual machine has the following features in security:

(1) It can quickly generate multiple trusted virtual isolated terminals for users without changing the existing network extension, and provide virtual isolated security for users with multiple security-grade services;

(2) By measuring the trusted state of the virtual machine monitor (VMM) in the boot phase to prevent the leakage of underlying information caused by the VMM being damaged;

(3) It can ensure that each user virtual machine can be trusted in the activated environment, and can avoid environment security threats caused by invading malicious code such as viruses and Trojans;

(4) It can control the applications that are loaded in different user virtual machine running phases as needed, that is, the applications specified with the trustworthy metric value satisfying the expectation are allowed to be loaded, so as to ensure the trusted and manageable operation in the running phase of the user virtual machine.

5 Conclusion

Starting from the bare-metal virtual machine, this paper proposes a new enhanced model based on the bare-metal virtual machine by researching the start-up and running mechanism of the bare-metal virtual machine, and designs the prototype system. The prototype system can establish a credible and safe operating environment for the bare-metal virtual machine system. It not only provides users with technical support for the safe deployment and use of the service system with multiple security levels, but also supports the development of virtualization technology and important information The construction of the system is of great significance.

References

1. Chang-xiang Shen, Network space security strategy thinking and Revelation [J], financial electronic, 2014(6):1-3.(In Chinese)
2. Chang-xiang Shen, Huan-guo Zhang, Huai-min Wang, research and development of trusted computing [J]. China Information Science (Chinese edition), 2010, 40 (2): 139-166. (In Chinese)
3. GB / T 29828-2013 Information security technology trusted computing specifications trusted connection architecture [M], Beijing: China Standard Press, 2013. (In Chinese).

A New Method of Command and Guide the Unbalanced Distribution of Target

GAO Yangyang, YU Minjian, YOU Hanghang and JI Huiming

Air Force Engineering University, Air Traffic Control and Navigation College, 710051 Xi'an, China

Abstract. In the process of air combat implementation, the real-time command and guidance of air pilot is the key element of air combat irreplaceable. In this paper, a multi-aircraft air combat command and guidance target allocation model is established based on the actual air combat situation. The unbalanced target allocation problem of multi-aircraft air combat command and guidance is studied by combining Hungarian algorithm and evolutionary Hungarian algorithm. Finally, two cases of increasing aircraft number and decreasing aircraft number are simulated and studied. Simulation results show that the method can shorten the computation time and ensure the overall command and guidance advantages of the target allocation scheme effectively.

1 Introduction

In the networked modern air combat of multi-aircraft cooperative confrontation, the guidance and allocation of enemy and fighter planes is an important link in the command and guidance of multi-aircraft air combat. Combine with that concrete reality of our army's command and guidance allocation problem, we mainly carry out one-to-one or many-to-one task allocation, so we can treat the command and guidance allocation problem as a target allocation problem.

Literature[1][2]has studied the target assignment of the United States air force unmanned aerial vehicle (UAV) cooperative operation. The problem of UAV cooperative control and task assignment is studied in reference [3]. At the domestic level, the Taiwan university of science and technology has conducted research on the distribution of weapons targets [4]; In the mainland, there are also many scholars on the target allocation problem has been studied, Lv Hongwei, etc from the air combat geometry situation, put forward to our overall geometric advantage of the largest target allocation algorithm, so as to give full play to the overall formation advantage [5]. Chen Jieyu et al. proposed a target allocation algorithm based on distributed auction mechanism, which can get the optimization scheme quickly [6]. [7] considering the advantages of simulated annealing algorithm, such as strong robustness, good universality and easy implementation, the simulated annealing algorithm is used to solve the weapon-target assignment problem.

Based on the practical consideration of multi-aircraft air combat, this paper divides the command and guidance assignment problem into small-scale assignment problem and large-scale assignment problem. Hungary algorithm and evolutionary Hungary algorithm are used for target assignment respectively, which not only ensures the accuracy of the assignment results, but also shortens the operation time.

2 Target assignment model

2.1 Concept of target assignment

Considering the number of aircraft, target allocation has two ways, one is balanced target allocation, the other is unbalanced target allocation.

2.1.1 Balanced target distribution

Balanced target distribution method refers to the air combat, both sides forces are consistent, and aircraft ability is equal. According to the battlefield situation, the enemy and our aircraft to " one - to - one" distribution mode, balanced distribution method is generally the original distribution method.

2.1.2 Unbalanced target distribution

In actual operations, the problem of balanced target allocation is only a relatively ideal situation. Usually, the number of aircraft on both sides is not equal. Unbalanced target allocation method is more complex than balanced target allocation method. During processing, the number of targets can be balanced by adding virtual targets, which can be transformed into balanced target allocation problem.

2.2 Multi - aircraft air combat command and guidance target assignment model

In multi-aircraft air combat command and guidance target allocation, only the number of enemy and our aircraft in air combat needs to be considered, so as to model the command and guidance target allocation. Suppose there are m enemy planes and n our planes in air combat, each of which can attack only one enemy plane at the same time. Our fighter plane is composed of cluster $W=\{W_i, i=1,2,\cdots,n\}$ and enemy target machine is composed of cluster $D=\{D_j, j=1,2,\cdots,m\}$. Among them, W_i represents our ith fighter, D_j represents the enemy's jth fighter, in order to achieve air combat victory, usually in the use of forces against the enemy forces constitute an advantage, so should meet $n \geq m$, the multi-aircraft air combat command guidance target allocation model can be described as follows:

$$z = \max \sum_{i=1}^{n} \sum_{j=1}^{m} C_{ij} X_{ij} \quad (1)$$

$$\sum_{i=1}^{n} X_{ij} = 1, j = 1,\cdots,m \quad (2)$$

$$\sum_{j=1}^{m} X_{ij} = 1 \text{ } or \text{ } 2, \quad i = 1,2,\cdots,n \quad (3)$$

Satisfy:

$$X(i,j) = \begin{cases} 1 \\ 0 \end{cases} \quad (4)$$

Among them, $C=(C_{ij})_{n\times m}$ is the evaluation matrix, and C_{ij} is the air combat command and guidance dominance value of the ith aircraft to the jth enemy aircraft. $X=(X_{ij})_{n\times m}$ is the solution matrix of target distribution, X_{ij} value of 1 said has been assigned the jth enemy aircraft to the ith aircraft, value of 0, said not to assign the jth enemy aircraft to the ith aircraft. Z is the objective function, said our fleet total command guidance advantage value, reflect our fleet total air combat situation advantage. An assignment solution X is given to maximize the commander-in-chief lead advantage value of Z . Constraints (2) and (3) respectively indicate that one of our aircraft can attack only one enemy aircraft and one enemy aircraft can be attacked by two of our aircraft at the same time, reflecting the characteristics of multi-aircraft air combat command and guidance target allocation.

3 Multi - objective task assignment based on Hungarian algorithm

3.1. Hungarian algorithm

Hungarian algorithm was proposed by Kuhn in 1955. it is a solution to assignment problem.

The general implementation steps of the Hungarian algorithm are as follows:

Step 1: Randomly number enemy and our fighter.

Step 2: Based on the attribute of enemy aircraft and the air combat situation of enemy and our aircraft, the situation assessment is carried out, and then the evaluation matrix is constructed.

Step 3: Linearly transform the evaluation matrix so that 0 elements appear in each column of each row. (Each line and column minus the smallest element).

Step 4: To obtain the optimal allocation result from the adjusted evaluation matrix.

3.2 Evolutionary Hungarian algorithm

Evolutionary Hungarian algorithm is proposed by Holland. Evolutionary algorithm can effectively solve the problem of target allocation, but the computational speed decreases significantly with the increase of dimension, which cannot meet the real-time requirements of battlefield. In this paper, we combine Hungarian algorithm with evolutionary Hungary to deal with large-scale target allocation. The algorithm flow is as follows:

Step 1: Judge the number of the fighter planes on both sides and randomly number the fighter planes. If the number of aircraft on both sides is not more than 4, can be directly calculated using simple Hungarian algorithm. If the number of aircraft is more than 4, it is considered as a large-scale target allocation problem, and the two algorithms are combined to deal with it.

Step 2: In dealing with large-scale target allocation problem, the evaluation matrix is firstly processed by Hungarian algorithm to obtain corresponding solution matrix, then several corresponding 0 – 1 matrices are randomly generated, and these 0 - 1 matrices and the obtained solution matrix are taken as the initial population of the algorithm.

Step 3: The fitness of each individual in the population is calculated, and replication and variation are carried out.

Step 4: Increasing the evolution algebra until the maximum evolution algebra is reached, and then selecting the individual in the obtained population, wherein the optimal individual is the required solution, and the corresponding target distribution result is the optimal distribution scheme.

3.3 Processing of dynamic target assignment

In air combat, the forces of both sides will change due to the damage or withdrawal of aircraft, so the target allocation will be a dynamic process.

3.3.1 The situation of aircraft out of the battlefield

When a fighter exits the battlefield, the rows or columns of the corresponding evaluation matrix should be deleted. however, in order to maintain the integrity of the evaluation matrix and enable the evaluation matrix to be used in the calculation and processing of the algorithm, according to the margin method proposed in this paper, the row or column element value of the corresponding evaluation matrix of the fighter exiting the battlefield is changed to 0, which not only ensures the integrity of the evaluation matrix, but also ensures that the changed element value does not affect the normal target allocation.

3.3.2 The situation of aircraft into the battlefield

When a fighter plane is newly added to a battlefield, the number of the fighter planes in the battlefield changes, and the evaluation matrix also changes accordingly, and the number of rows and columns of the evaluation matrix is not equal at this time, and the edge mending method is adopted for processing.

4 Simulation analysis

4.1. Increased number of aircraft

In air combat, there are three kinds of initial situations of the number of fighter planes on both sides: the number of our aircraft is dominant, the number of enemy aircraft is dominant and the number of both sides is equal. But in actual operations, in order to ensure the operational effect, often won't appear under the initial situation of our number is less than the number of enemy aircraft, our number is dominant, the number of enemy aircraft after increase can be converted into a balanced target allocation problem, in this paper, we discuss the number of both sides is equal. Assume that the corresponding evaluation matrix is $T_{4\times4}$.

$$T_{4\times4}=\begin{bmatrix}245 & 345 & 578 & 335\\ 567 & 680 & 123 & 280\\ 467 & 134 & 232 & 167\\ 225 & 115 & 245 & 496\end{bmatrix} \tag{5}$$

The Hungarian algorithm is use for calculation and processing, and that result matrix is as follow:

$$T_{4\times4}=\begin{bmatrix}0 & 0 & 1 & 0\\ 0 & 1 & 0 & 0\\ 1 & 0 & 0 & 0\\ 0 & 0 & 0 & 1\end{bmatrix} \tag{6}$$

The corresponding allocation results are as follows: our aircraft No.1 attacked enemy aircraft No.3; Our aircraft No. 2 attack enemy aircraft No. 2; Our aircraft No. 3 attack enemy aircraft No. 1; Our aircraft No. 4 attacked enemy aircraft No. 4.

If suddenly increase aircraft, according to the edge method to change the evaluation matrix. In this initial condition, if the enemy add a fighter, through the analysis of the battlefield situation, get the new fighter threat value, the corresponding after the evaluation matrix to add a column of the aircraft threat value, at the same time need to add the corresponding row below the evaluation matrix for " edge", the principle is that does not affect the normal target distribution. The modified evaluation matrix is:

$$T_{5\times5}=\begin{bmatrix}245 & 345 & 578 & 335 & 99\\ 567 & 680 & 123 & 280 & 322\\ 467 & 134 & 232 & 167 & 431\\ 225 & 115 & 245 & 496 & 506\\ 0 & 0 & 0 & 0 & 0\end{bmatrix} \tag{7}$$

The result matrix is:

$$T_{5\times5}=\begin{bmatrix}0 & 0 & 1 & 0 & 0\\ 0 & 1 & 0 & 0 & 0\\ 1 & 0 & 0 & 0 & 0\\ 0 & 0 & 0 & 0 & 1\\ 0 & 0 & 0 & 1 & 0\end{bmatrix} \tag{8}$$

The corresponding allocation results are as follows: our aircraft No.1 attacked enemy aircraft No.3; Our aircraft No. 2 attack enemy aircraft No. 2; Our aircraft No. 3 attack enemy aircraft No. 1; Our aircraft No. 4 attack enemy aircraft No. 5; Our aircraft No. 5 attack enemy aircraft No. 4. As shown in figure 1:

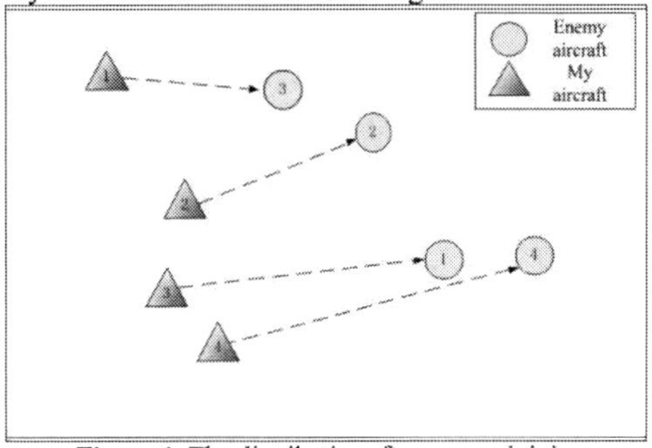

Figure 1. The distribution after enemy joining

4.2. Reduced number of aircraft

In the discussion of the above initial conditions, it is assumed that the corresponding evaluation matrix is $T_{4\times4}$.

$$T_{4\times4}=\begin{bmatrix}245 & 345 & 578 & 335\\ 567 & 680 & 123 & 280\\ 467 & 134 & 232 & 167\\ 225 & 115 & 245 & 496\end{bmatrix} \tag{9}$$

The result matrix is:

$$T_{4\times4}=\begin{bmatrix}0 & 0 & 1 & 0\\ 0 & 1 & 0 & 0\\ 1 & 0 & 0 & 0\\ 0 & 0 & 0 & 1\end{bmatrix} \tag{10}$$

The corresponding allocation results are as follows: our aircraft No.1 attacked enemy aircraft No.3; Our aircraft No. 2 attack enemy aircraft No. 2; Our aircraft No. 3 attack enemy aircraft No. 1; Our aircraft No. 4 attacked enemy aircraft No. 4.

When the enemy aircraft No. 3 out of the battlefield, using the edge method need to modify the third column is as follows:

$$T'_{4\times4} = \begin{bmatrix} 245 & 345 & 0 & 335 \\ 567 & 680 & 0 & 280 \\ 467 & 134 & 0 & 167 \\ 225 & 115 & 0 & 496 \end{bmatrix} \quad (11)$$

The result matrix is:

$$T'_{4\times4} = \begin{bmatrix} 0 & 0 & 1 & 0 \\ 0 & 1 & 0 & 0 \\ 1 & 0 & 0 & 0 \\ 0 & 0 & 0 & 1 \end{bmatrix} \quad (12)$$

The corresponding allocation results are as follows: our aircraft No.1 attacked enemy aircraft No.3; Our aircraft No. 2 attack enemy aircraft No. 2; Our aircraft No. 3 attack enemy aircraft No. 1; Our aircraft No. 4 attacked enemy aircraft No. 4.

As shown in figure 2:

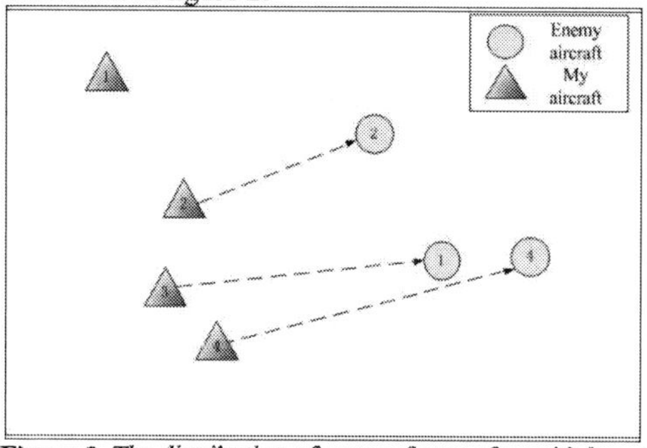

Figure 2. The distribution of enemy forces after withdrawal

In the case of Figure 2, you can choose to match the aircraft No. 2 attack on enemy aircraft No. 2 with my aircraft No.1, and the redistribution is shown in Figure 3.

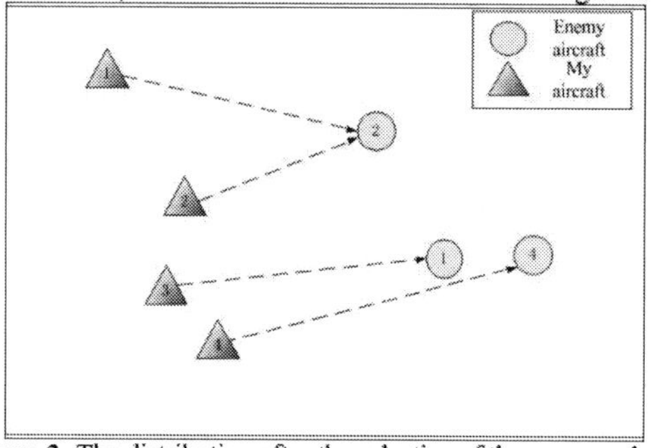

Figure 3. The distribution after the selection of the commander

5 Conclusion

Based on Hungarian algorithm, this paper studies the problem of command and guidance target assignment. Hungary algorithm is used to discuss the unbalanced target allocation method and the dynamic reallocation method after the forces of both sides withdraw from the battlefield under the condition of the battlefield, and the simulation results are given. The overall command and guidance advantages of the target allocation scheme are effectively ensured.

References

1. KINGSTON D B, RASMUSSEN S J, MEARS M J. Base defense using a task assignment framework ［C］/ /AIAA Guidance, Navigation and Control Conference, Chicago, Illinois, doi: 10.2514/6.2009-6209.
2. ANDERSSON K, KAMINER I, JONES K D, et al. Cooperating UAVs using thermal lift to extend endurance ［C］/ /AIAA Infotech @ Aerospace Conference, Washington doi:10.2514 /6. 2009-2043.
3. NIGAM N, KROO I. Control and design of multiple unmanned air vehicles for a persistent surveillance task ［C］/ /12th AIAA /ISSMO Multidisciplinary Analysis and Optimization Conference, Victoria, British Columbia Canada doi:10. 2514/6. 2008-5913.
4. LEE Z J, SU S F, LEE C Y. Efficiently solving general weapon-target assignment problem by genetic algorithms with greedy eugenics ［J］. IEEE Transactions on Systems, Man, and Cybernetics, Part B: Cybernetics, 2003,33(1) :113-121.
5. LI X M，LI J F,et al. Method of Static Precision Corrected of Carrier Radar Based on CCD-Laser-Theodolite[J]. Journal of Naval Aeronautical and Astronautical University,2008,23(1): 59-61.
6. Wu P, Liang Q. Simulated Annealing Algorithm for Weapon- Target Assignment Problem[J].Computer & Digital Engineering, 2013,41(2):201-203.
7. CHEN J ,YAO P Y. Multi-UAV Decentral ized Coopertative Dynamic Target Assignment Method[J]. Journal of Air Force Engineering University,2014,15(6):11-16.
8. WANG Z,LIU Z Y, PEI R,et al. State Estimation For Uncertain System By Moving Horizon Strategy[J]. Proceedings of the CSEE,2003,23(9):99-102.
9. ZHAO H Y, GAO X Q,C H.Application of moving horizon estmiation in three tank system[J].Journal of Jilin University (Information Science Edition), 2004,22 (4): 298- 302.
10. KANG L, ZHAO C X, GUO J H. Path Planning Based on Fuzzy Rolling Rapidly exploring Random Tree for Mobile Robot[J].Journal of Nan jing University of Science and Technology (Natural Science),2010,35(5):642-648.
11. HUO A Q,DAI C,WANG Y L, et al. Predictive control in stabilized platform based on augmented non-minimalstate - spacemethod.ComputerEngineeringandApplications, 2013,49,(17):239-244.

RBF Neural Network Control for USV with Input Saturation

Hua Deng, Renqiang Wang, Jingdong Li, Dawei Chen, Jianming Sun, Yue Zhao, Jiabao Du

College of Navigation, Jiangsu Maritime Institute, Nanjing 211170, PR China

Abstract. Intelligent control for USV with input saturation based on RBF network was proposed. Firstly, sliding surfaces with integral were designed on the basis of the sliding mode variable structure control technology. Secondly, RBF network was applied to approximate compensate the input saturation of system, and which was optimized by Genetic Algorithms. Finally, the control algorithm for USV was deduced by backstepping method with Lyapunov theory on the basis of sliding mode control. Relevant simulations show the control method is available for USV motion control.

1 Introduction

Unmanned Surface Vehicle (USV) due to the advantages of small size, mobility and other characteristic has become an important development trend of the future of the ship and which is mainly used to implement the dangerous task. So, it is playing increasing roles in commercial, scientific and military applications [1]. As the disturbances of wind, wave, current and other marine environment, the course of USV will inevitably deviate from the original given course when navigating at sea. While course control is a prerequisite to solve the trajectory tracking, autonomous navigation and collision avoidance and other issues. So, Steering control is one of the important research topics in the field of motion control [2] for USV.

At present, the motion control of most ships is on the basis of PID which has the advantages of simple algorithm structure and easy project realization, but its control parameters entirely rely on artificial experience judgments, therefore, it can only achieve limited intermittent adjustment with poor adaptive ability. A series of control theories and methods have been applied in USV heading control with the rapid development and progress of computer technology, such as Lyapunov stability theory and method [3], backstepping control algorithm [4,5], sliding mode control theory and technology [6-8], intelligent control theory and technology represented by fuzzy algorithm [9-12] and neural network [13-15].

In this paper, a kind of RBF network control algorithm for USV is proposed on the basis of Nomoto mathematical model and combination of integral backstepping control and sliding mode control.

2 Problem description

2.1. USV plane motion model

In actual straight-line voyages, ship movements usually exhibit a non-linear state. Therefore, in the mathematical model of plane motion, the nonlinear model for USV is described as following with influence of interference.

$$\begin{cases} \dot{x}_1 = x_2 \\ \dot{x}_2 = f(x) + g \cdot u + dt \\ y = x_1 \end{cases} \tag{1}$$

Where, x_1 and x_2 are the state variables of control system, y is the actual course, $f(x)$ is the uncertain mathematical function of the system, and $f(x) = -\frac{1}{T}H(\dot{\varphi})$, $H(\dot{\varphi})$ is unknown nonlinear mathematical function, $H(\dot{\varphi}) = \dot{\varphi} + \alpha\dot{\varphi}^3$, φ is actual course too, α is the constant to be determined. g is the gain of control input, and $g = K/T$. u is control input representing rudder angle. dt is external interference. K is the index of ship's turning performance and T is the index of ship's followability. These parameters can be obtained through ship handling experiment.

2.2. Anti-saturation compensation based on RBF network

It is well known that the USV is generally drived by rudder servo system. Thus, the actual characteristics of the rudder servo system, in general, the actual output of the rudder angle is less than 35 °, must be taken into account in actual navigation with large-scale motor. To solve this problem, in this paper, the anti-saturation theory is cited first, and an online approximation and effective compensation is achieved by RBF neural network.

Anti-saturation compensation for rudder servo system is described as follows, the maximum control input value is introduced as u_{max}, and maximum value is 35°. In addition, it is constructed that $\delta = u - v$ and $u = sat(v)$. Therefore, the control input restricted mathematical formula $sat(v)$ was shown in equation (2).

$$sat(v) = \begin{cases} u_{max} & v > u_{max} \\ v & |v| \le u_{max} \\ -u_{max} & v < -u_{max} \end{cases} \tag{2}$$

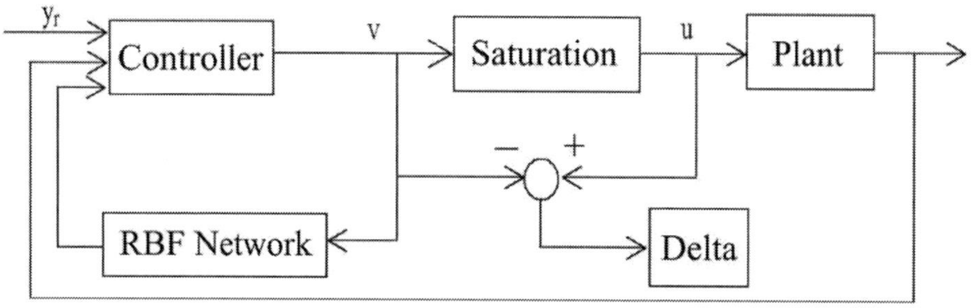

Fig. 1. Closed-loop control system with saturation

In paper, the RBF neural network is designed to approximate rudder output δ so as to realize effective compensation, where closed-loop control system diagram for USV is shown in figure 1 on the basis of RBF neural network compensation.

It is described that the mathematical model of RBF network is shown in equation (3).

$$\begin{cases} h_j = \exp\left(\dfrac{\|\mathbf{x} - \mathbf{c}_j\|^2}{2b_j^2} \right) \\ \delta = \mathbf{W}^{*T} \cdot \mathbf{h}(\mathbf{x}) + \varepsilon \end{cases} \tag{3}$$

Where, $\mathbf{x}$ is the input of RBF network, j is a hidden layer of the network the first network input, $\mathbf{h} = [h_j]^T$ is high Gaussian function, $\mathbf{W}^*$ is the ideal for weighting value, ε is ideal for neural network approximation error, and $\varepsilon \le \varepsilon_{max}$.

According to figure 1, the RBF network input is $x = v$, the network output $\hat{\delta}$ is obtained by equation (4).

$$\begin{cases} \hat{\delta} = \hat{\mathbf{W}}^{\mathrm{T}} \mathbf{h}(\mathbf{x}) \\ \tilde{\delta} = \delta - \hat{\delta} = -\tilde{\mathbf{W}}^{\mathrm{T}} \mathbf{h}(\mathbf{x}) + \varepsilon \end{cases} \tag{4}$$

Where, $\tilde{\mathbf{W}} = \hat{\mathbf{W}} - \mathbf{W}^*$.

2.3 RBF Neural Network Optimization

RBF neural network has a universal property of approximation [16], while its ability of approximation depends entirely on network performance parameters, i.e., output weights w_i, width b_i and the number of hidden center units c_i. However, it is very difficult to determine these parameters in advance. In view of that, GA is uesd to optimize RBF neural network, which is able to approach the internal uncertainties of ship model and external disturbance, so as to achieve the best performance of approaching. The process of RBF neural network parameters optimization is shown in Figure 2.

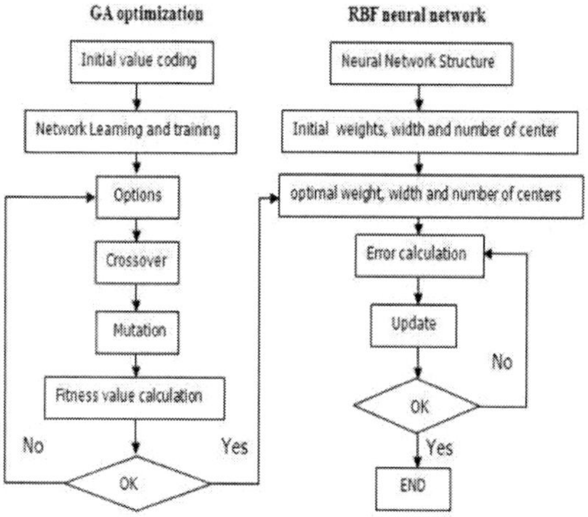

Fig. 2. RBF neural network optimization with GA

3 Intelligent Steering control design based on backstepping

The designing process of control law contains the following steps.

Firstly, according to the characteristics of the system (1), the sliding surfaces are defined as following:

$$\begin{cases} \dot{\xi} = z_2 \\ z_1 = x_1 - \varphi_r \\ z_2 = x_2 - \sigma_1 \end{cases} \tag{5}$$

Where, φ_r is expected course, σ_1 and σ_2 are virtual stabilization function, z_1 is heading error, ξ is the integral term which is added in order to eliminate the static error.

Secondly, constructing the Lyapunov function to determine the asymptotic stability of control system.

According to equation (5), it is obtained equation (6).

$$\dot{z}_1 = z_2 + \sigma_1 - \dot{\varphi}_r \tag{6}$$

The virtual stabilization σ_1 is set as equation (7)

$$\sigma_1 = -k_1 z_1 + \dot{\varphi}_r \tag{7}$$

Therefore, it can be obtained equation (8).

$$\dot{\sigma}_1 = -k_1\left(-k_1 z_1 + z_2\right) + \ddot{\varphi}_r \tag{8}$$

Similarly, it is obtained equation (9) and equation (10) according to equation (5).

$$\dot{z}_1 = -k_1 z_1 + z_2 \tag{9}$$

$$\dot{z}_2 = \dot{x}_2 - \dot{\sigma}_1 = f(x) + gu + dt - \dot{\sigma}_1 \tag{10}$$

And, the first Lyapunov function is constructed as equation (11).

$$V_1 = \frac{1}{2}z_1^2 + \frac{1}{2}z_2^2 \tag{11}$$

It can be inferred equation (12) by equation (11).

$$\dot{V}_2 = -k_1 z_1^2 + z_1 z_2 + z_2\left[f(x) + gu + dt - \dot{\sigma}_1\right] \tag{12}$$

The second Lyapunov function is constructed as equation (13).

$$V_3 = V_2 + \frac{\lambda}{2}\xi^2 \tag{13}$$

Similarly, it can be inferred equation (14) by equation (13).

$$\begin{aligned}\dot{V}_3 &= \dot{V}_2 + \lambda\xi\dot{\xi} \\ &= -k_1 z_1^2 + z_2\left[z_1 + f(x) + gu + dt - \dot{\sigma}_1 + \lambda\xi\right]\end{aligned} \tag{14}$$

Thirdly, according to figure 2, it is known that $u = v + \delta$, which is taken into equation (14). Therefore, it can be obtainedequation (15).

$$\dot{V}_3 = -k_1 z_1^2 + z_2\left[z_1 + f(x) + g(v+\delta) + dt - \dot{\sigma}_1 + \lambda\xi\right] \tag{15}$$

The following control law with damping law [9] is designed by the RBF network.

$$v = \frac{1}{g}\left[-z_1 - f(x) - (k_2 + \eta)z_2 + \dot{\sigma}_1 - \lambda\xi\right] - \hat{\delta} \tag{16}$$

It can be inferred equation (17), according to equation (15) and equation (16).

$$\dot{V}_3 = -k_1 z_1^2 - k_2 z_2^2 - \eta z_2^2 + z_2 g\tilde{\delta} + z_2 dt \tag{17}$$

Finally, the total Lyapunov function is constructed as equation (18).

$$L = V_3 + \frac{1}{2}\gamma\tilde{\mathbf{W}}^T\tilde{\mathbf{W}} \tag{18}$$

Where, $\gamma > 0$.

Therefore, it can be inferred equation (19) by equation (18)

$$\begin{aligned}\dot{L} &= -k_1 z_1^2 - k_2 z_2^2 - \eta z_2^2 + z_2 g\tilde{\delta} + z_2 dt + \gamma\tilde{W}^T\dot{\hat{W}} \\ &= -k_1 z_1^2 - k_2 z_2^2 - \eta z_2^2 + \tilde{W}^T(\gamma\dot{\hat{W}} - \mathbf{h}z_2) + \left[d(t) + \varepsilon + g\tilde{\delta}\right]z_2\end{aligned} \tag{19}$$

In order to ensure system asymptotically stable, the following adaptive law is designed.

$$\dot{\hat{W}} = \frac{1}{\gamma}\mathbf{h}z_2 \tag{20}$$

It is easy obtained formula (21) by theorem "$xy \le \eta x^2 + \frac{1}{4\eta}y^2$".

$$\varpi z_2 \le \eta z_2^2 + \frac{1}{4\eta}\|\varpi\|_\infty^2 \tag{21}$$

Where, $\varpi = d(t) + \varepsilon + g\tilde{\delta}$.

Therefore, the formula (22) can be acquired .

$$\begin{aligned}\dot{L} &= -k_1 z_1^2 - k_2 z_2^2 - \eta z_2^2 + \varpi z_2 \\ &\le -k_1 z_1^2 - k_2 z_2^2 + \frac{1}{4\eta}\|\varpi\|_\infty^2 \\ &\le -k_2 z_2^2 + \frac{1}{4\eta}\|\varpi\|_\infty^2\end{aligned} \tag{22}$$

In order to ensure that the system (1) is asymptotically stable, it is only need to prove the establishment which is $k_2 z_2^2 \ge \frac{1}{4\eta}\|\omega\|_\infty^2$. Proof slightly, see the literature [4].

So, the entire control system is asymptotically stable.

4 Simulation

4.1. Simulated USV model

The USV "Lanxin" was adopted in simulation, and the parameters of planar motion mathematical model were listed as table 1. So, the parameters of Nomoto model of "Lanxin" were $K = 1.502$, $T = 0.905$, $\alpha_1 = 0.001$, $\alpha_{i(i=1,2,3\cdots)} = 0$, when its speed is 8.5 knots.

Table 1. The "LanXin" USV Parameters

Length Between Perpendiculars	7.02 m
Breadth	2.60 m
Speed	<=35 kn
Draft (full load)	0.32 m
Block Coefficient	0.6976
Displacement (full load)	2.73 m³
Rudder Area	0.2091 m²
Distance Between Barycenter and Center	0.35 m

At the same time, the model of rudder servo system which was described as equation (23) was added into nonlinear Nomoto model of "Lanxin".

$$\dot{\delta} = -\frac{1}{T_r}\delta + \frac{1}{T_r}\delta_r \tag{23}$$

Where, T_r is time constant, and the value of which is generally about 0.2 seconds. δ_r is executed rudder angle. δ is real-time rudder angle, and the restriction of which was shown in equation (24).

 https://doi.org/10.1051/matecconf/201821403002

$$\delta = \begin{cases} 35^\circ & \delta > 35^\circ \\ \delta & |\delta| \le 35^\circ \\ -35^\circ & \delta < -35^\circ \end{cases} \qquad (24)$$

In addition, it is considered that ship's movement is greatly affected by wind and wave. And the model can be replaced by a transfer function of second-order wave and a white noise [17].

4.2. Condition setting

The ideal angle command is $x_d = \sin(t)$. In order to show that the control system compensates the control input limited capacity, a larger initial error is used, and the initial vector of the system is $\begin{bmatrix} 10 & 0 \end{bmatrix}$. RBF network structure is adopted as figure 1, and network input is $x = v$, according to the actual input range of the network to design Gaussian function of the parameters, taking $C_i = \begin{bmatrix} -1.0 & -0.5 & 0 & 0.5 & 1.0 \end{bmatrix}$ and $b_i = 5$, the network initial value is zero. Using the control law equation (19) and adaptive law equation (24), where $c = 5$, $\gamma = 10$. In the sliding mode control, the saturation function is used instead of the switching function, and the boundary layer thickness Δ is 0.02.

4.3. Simulation and analysis

Simulation of USV course tracking control was carried out on the basis of Matlab/Simulink platform, and the results were shown in figure 3, figure4 and figure5.

The output response of angle tracking and speed tracking of USV was depictted in figure 3. And from two pictures, it can be seen that course angle reaches the set value with no overshoot.

And the output response of restricted front and rear control inputs was described in figure 4. And from this picture, it can be seen that an internal auxiliary compensation mechanism designed with anti-saturation control technique can make control input quickly quit saturation.

And also the output response of rudder angle and approximation was bewrited in figure 5. And from this picture, it can be seen that an internal auxiliary compensation mechanism on the basis of RBF network has effective online approximation, which is proposed by anti-saturation control technique.

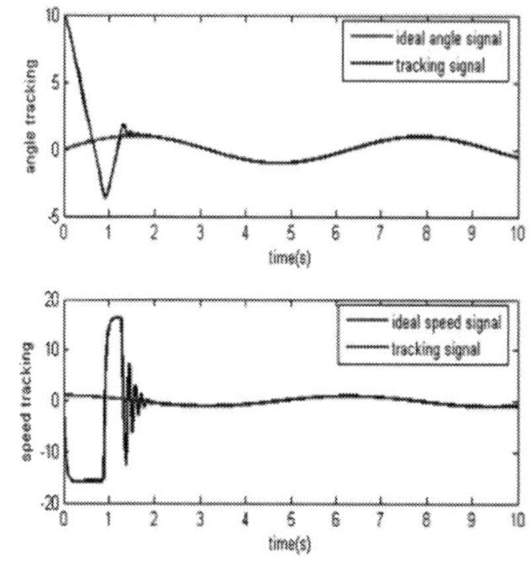

Fig. 3. Position and speed tracking.

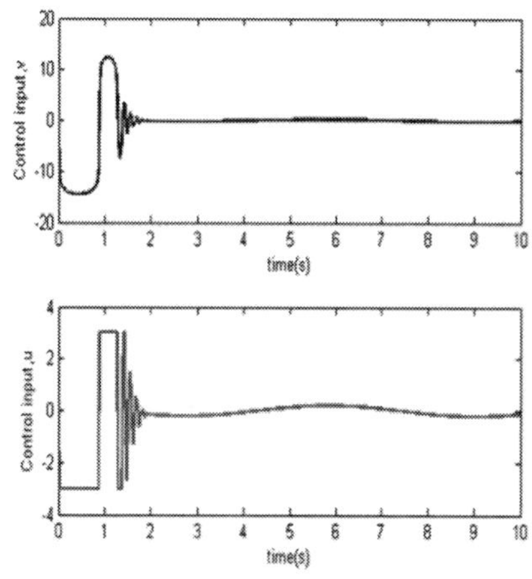

Fig. 4. Restricted front and rear control inputs v and u.

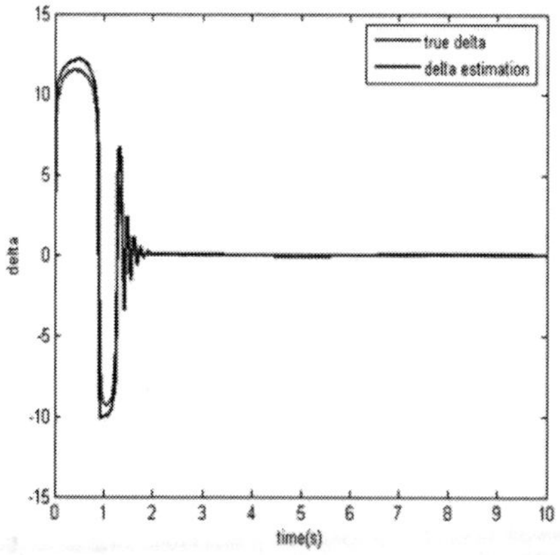

Fig. 5. Controls the input limit value δ and the approximation.

5 Conclusion

Intelligent steering control for USV with input saturation was proposed based on RBF network. It is shown that an internal auxiliary compensation mechanism designed with anti-saturation control technique can make control input quickly quit saturation, and RBF neural network optimized by GA has effective online approximation. Therefore, the intelligent control algorithm designed in this paper is available for USV.

Acknowledgment

This work was supported by Natural Science Research Project of Universities in Jiangsu Province under Grant No.18KJB580003, the Foundation of Jiangsu Maritime Institute under Grant No.XR1501, and Top-notch Academic Programs of Jiangsu Higher Education Institutions under Grant No. PPZY2015B177.

References

1. S. Bennett, Nicolas Minorsky and the automatic steering of ships, IEEE Control Systems Magazine, 4, 1984, 10-15.
2. S. Kumarawadu, K. Kumara, On the speed control for automated surface vessel operation, in Proceedings of the 2007 third international conference on information and automation for sustainability, Melbourne, Australia, 2007, 135-140.
3. B. Volker, Unmanned Surface Vehicles - A Survey, Skibsteknisk Selskab, Copenhagen, Denmark, 2008.
4. Benaskeur A. R, Desbines A, Backstepping based adaptive PID contro, Proceedings of Control Theory and Applications, 149, 2002, 54-59.
5. Y. Pan, D. Huang, Z. Sun, Backstepping adaptive fuzzy control for track-keeping of under-actuated surface vessels, Control Theory and Applications, 7, 2011, 907-914.
6. Wang Renqiang, Gong Jianyun, Zhao Yuelin, Neural Network Sliding Mode Control under New Reaching Law and Application, Proceedings of Advanced Information Technology, Electronic and Automation Control Conference, 12, 2015, 911-917.
7. Renqiang Wang, Yuelin Zhao, Keyin Miao, Application of Neural Network Minimum Parameter Learning Algorithm in Ship's Heading Tracking Control, Proceedings of 9th International Symposium on Computational Intelligence and Design, 12, 2016, 135-139.
8. K. Do, Z. Jiang, J. Pan, Under-actuated ship global tracking under relaxed conditions, IEEE Trans. on Automatic Control, 47, 2002, 1592 – 1536.
9. Wang Ning, Lv Shuailin, Liu Zhongzhong, Global finite-time heading control of surface vehicles, Neurocomputing, 175, 2016, 662-666.
10. X. Bao, K. Nonami, Z. Yu, Combined yaw and roll control of an autonomous boat, Proceedings of the 2009 IEEE International Conference on Robotics and Automation, Kobe, Japan, 2009, 188-193.
11. T. Li, Y. Yang, B. Hon, Robust adaptive fuzzy design for ships track-keeping control, Control Theory and Applications, 03, 2007, 445-448.
12. L. Yuan, H. Wu, Multiple sliding mode adaptive fuzzy controller for nonlinear marine autopilot systems, CAAI Transactions on Intelligent Systems, 04, 2010, 308-312.
13. Guanshan Hu, Yingbing Zhou, Application of Fuzzy Neural Network in Ship Course Control, Applied Mechanics and Materials, 1511, 2012, 309-315.
14. Xin Wang, Tieshan Li, Weilin Luo, Direct Adaptive Neural Network Control for a Class of Ship Course Uncertain Discrete-time Nonlinear Systems, Marine Engineering Frontiers, 37, 2013, 42-48.
15. Yang Wang, Chen Guo, Fuchun Sun, Dynamic neural fuzzified adaptive control of ship course with parametric modelling uncertainties, Int. J. of Modelling, Identification and Control, 13, 2015, 251-258.
16. Wang Renqiang, Zhao Yuelin, Sun Jianming, Application of Optimized RBF Neural Network in Ship's Autopilot Design, Proceedings of IEEE Advanced Information Management, Communicates, Electronic and Automation Control Conference, 10, 2016, 1642-1646.
17. FAN Yunsheng, SUN Xiaojie, WANG Guofeng, GUO Chen, On Fuzzy Self-adaptive PID Control for USV Course, Proceedings of the 34th Chinese Control Conference, 7, 2015, 8472-8478.

ADRC Attitude Controller Design for Hypersonic Vehicle based on MIMO-ESO

JianHua Wen[1], ChangMao Qin[1] and Xin Zhang[1]

[1]*Senior Engineer,China Academy of Launch Vehicle Technology,Mailbox 9200 sub box 1-1,Beijing,100076, China*

Abstract. For the hypersonic vehicle nonlinear attitude mode in reentry process with a strong coupling, aerodynamic parameter perturbations and non-deterministic, combine extended state observer and nonlinear law state error feedback, design the hypersonic vehicle MIMO-ESO ADRC attitude controller. Put interference such as uncertainty, coupling and parameter perturbations as "the sum of interference" ,use the extended state observer to estimate and dynamic feedback compensation, use nonlinear law state error feedback to inhibit residual of compensation. ADRC controller is charged without a precise model of vehicle , and without precise perturbation boundaries of aerodynamic parameters.Simulation results show that the MIMO-ESO ADRC attitude controller can overcome the impact of large-scale perturbations of interference and aerodynamic parameters, have good dynamic qualities and tracking capabilities, also have strong robustness.

1 Introduction

Hypersonic vehicle has important military significance. It has become a research hotspot in various countries in recent years[1]. The controller design has been more and more difficult due to the complex nonlinearity, the strong coupling between control channels and the uncertainties of aerodynamic parameters in the non-power re-entry process. In existing methods such as robust control[2,3] and adaptive control[4,5], the linearized model of the controlled system contains complex high-order derivative function, which is not convenient for practical application of engineering. Sliding mode control[6,7] compensates the influence of uncertainty by designing the virtual control quantity, but the sliding mode control is based on the upper bound of uncertainty and requires that its bound is known or the known function of the state variables, which is difficult to predict in practical applications.

The active disturbance rejection controller (ADRC)[8] is not dependent on the system model, but is a method that relies on the process error to eliminate the error. The extended state observer (ESO) estimates the "summation perturbation" to obtain the real-time effect amount of object model's internal and external disturbances, and with the real-time dynamic feedback compensation system is successfully linearized. In the meantime, nonlinear feedback control law (NLSEF) are used to suppress compensation residuals and improve control performance[9]. After more than ten years of development, ADRC technology has gradually matured. In 2013, Texas Instruments released a global motion control chip with ADRC as the core, further demonstrating the great potential of ADRC technology in engineering applications[10,11].

In this paper, the MIMO-ESO auto disturbance-rejection attitude controller is designed based on the nonlinear attitude model of the hypersonic vehicle's non-power re-entry. Uncertain items, unmodeled dynamics, coupling effects, parameter perturbations, and interference influences are considered as "summation interferences." ESO are used for estimation and dynamic feedback compensation, and NLSEF is used to suppress the compensation residuals. Depending on the characteristics that the auto-disturbance rejection is not relying on the model, the problem of complex linearization model and sliding model control requires perturbation is successfully solved. Selecting parameters according to ESO stability conditions can obtain good dynamic quality and tracking performance, and can also overcome the influence of interference and aerodynamic parameters over a wide range of perturbations, and has strong robustness.

2 Hypersonic Vehicle Modeling

The non-power re-entry attitude model of a hypersonic vehicle is a nonlinear system with uncertain items as follows:

$$\dot{x} = f(x) + g(x)\delta \quad (1)$$

As $x = [\alpha \quad \beta \quad \gamma_c \quad \omega_x \quad \omega_y \quad \omega_z]^T$, $\delta = [\delta_e \quad \delta_a \quad \delta_r]^T$.

Among them:

$$f(x) = [f_\alpha \quad f_\beta \quad f_{\gamma_c} \quad f_{\omega_x} \quad f_{\omega_y} \quad f_{\omega_z}]^T$$

$$f_\alpha = \omega_z - \omega_x \cos\alpha \tan\beta + \omega_y \sin\alpha \tan\beta$$
$$-\frac{1}{mV\cos\beta}[qSC_{Y,\alpha} + mg\cos\theta\cos\gamma_c]$$
$$f_\beta = \omega_x \sin\alpha + \omega_y \cos\alpha$$
$$+\frac{1}{mV}[qSC_{Z,\beta}\beta - mg\cos\theta\sin\gamma_c]$$
$$f_{\gamma_c} = \omega_x\frac{\cos\alpha}{\cos\beta} - \omega_y\frac{\sin\alpha}{\cos\beta} + \frac{g}{V}\cos\theta\cos\gamma_c\tan\beta$$
$$+\frac{1}{mV}[qSC_{Y,\alpha}(\sin\theta\sin\gamma_c + \tan\beta) + qSC_{Z,\beta}\beta\sin\theta\cos\gamma_c]$$
$$f_{\omega_x} = \frac{(I_y - I_z)}{I_x}\omega_y\omega_z + \frac{1}{I_x}qbS(C_{l,\beta}\beta$$
$$+C_{l,\omega_x}\frac{\omega_x b}{2V} + C_{l,\omega_y}\frac{\omega_y b}{2V})$$
$$f_{\omega_y} = \frac{(I_z - I_x)}{I_y}\omega_x\omega_z + \frac{1}{I_y}qbS(C_{m,\beta}\beta$$
$$+C_{m,\omega_x}\frac{\omega_x b}{2V} + C_{m,\omega_y}\frac{\omega_y b}{2V}) + \frac{1}{I_y}X_{cg}qSC_{Z,\beta}\beta$$
$$f_{\omega_z} = \frac{(I_x - I_y)}{I_z}\omega_y\omega_x + \frac{1}{I_z}qcS(C_{n,\alpha}$$
$$+C_{n,\omega_z}\frac{\omega_z c}{2V}) + \frac{1}{I_z}X_{cg}qS(C_{X,\alpha}\sin\alpha + C_{Y,\alpha}\cos\alpha)$$
$$g(x) = \begin{bmatrix} g_{\alpha,\delta_e} & g_{\alpha,\delta_a} & 0 \\ g_{\beta,\delta_e} & g_{\beta,\delta_a} & g_{\beta,\delta_r} \\ g_{\gamma_c,\delta_e} & g_{\gamma_c,\delta_a} & g_{\gamma_c,\delta_r} \\ g_{\omega_x,\delta_e} & g_{\omega_x,\delta_a} & g_{\omega_x,\delta_r} \\ g_{\omega_y,\delta_e} & g_{\omega_y,\delta_a} & g_{\omega_y,\delta_r} \\ g_{\omega_z,\delta_e} & g_{\omega_z,\delta_a} & g_{\omega_z,\delta_r} \end{bmatrix}$$
$$g_{\alpha,\delta_e} = -\frac{qSC_{Y,\delta_e}}{mV\cos\beta},\ g_{\alpha,\delta_a} = -\frac{qSC_{Y,\delta_a}}{mV\cos\beta},\ g_{\beta,\delta_e} = \frac{qSC_{Z,\delta_e}}{mV}$$
$$g_{\beta,\delta_a} = \frac{qSC_{Z,\delta_a}}{mV},\ g_{\beta,\delta_r} = \frac{qSC_{Z,\delta_r}}{mV}$$
$$g_{\gamma c,\delta_e} = \frac{qS[C_{Y,\delta_e}(\sin\theta\sin\gamma_c + \tan\beta) + C_{Z,\delta_e}\sin\theta\cos\gamma_c]}{mV}$$
$$g_{\gamma c,\delta_a} = \frac{qS[C_{Y,\delta_a}(\sin\theta\sin\gamma_c + \tan\beta) + C_{Z,\delta_a}\sin\theta\cos\gamma_c]}{mV}$$
$$g_{\gamma c,\delta_r} = \frac{qSC_{Z,\delta_r}\sin\theta\cos\gamma_c}{mV},\ g_{\omega_x,\delta_e} = \frac{qbSC_{l,\delta_e}}{I_x}$$
$$g_{\omega_x,\delta_a} = \frac{qbSC_{l,\delta_a}}{I_x},\ g_{\omega_x,\delta_r} = \frac{qbSC_{l,\delta_r}}{I_x}$$
$$g_{\omega_y,\delta_e} = \frac{qbSC_{m,\delta_e}}{I_y} + \frac{X_{cg}qSC_{Z,\delta_e}}{I_y}$$
$$g_{\omega_y,\delta_a} = \frac{qbSC_{m,\delta_a}}{I_y} + \frac{X_{cg}qSC_{Z,\delta_a}}{I_y}$$
$$g_{\omega_y,\delta_r} = \frac{qbSC_{m,\delta_r}}{I_y} + \frac{X_{cg}qSC_{Z,\delta_r}}{I_y}$$
$$g_{\omega_z,\delta_e} = \frac{qcSC_{n,\delta_e}}{I_z} + \frac{X_{cg}qS(\sin\alpha C_{X,\delta_e} + \cos\alpha C_{Y,\delta_e})}{I_z}$$
$$g_{\omega_z,\delta_a} = \frac{qcSC_{n,\delta_a}}{I_z} + \frac{X_{cg}qS(\sin\alpha C_{X,\delta_a} + \cos\alpha C_{Y,\delta_a})}{I_z}$$
$$g_{\omega_z,\delta_r} = \frac{qcSC_{n,\delta_r}}{I_z} + \frac{X_{cg}qS\sin\alpha C_{X,\delta_r}}{I_z}$$

Among the above symbols, $g_{i,j}$ is an aerodynamic parameter item. Its specific parameters are described in reference [12]. Different from ordinary low-speed vehicles, each aerodynamic parameter term is a function of angle of attack and Mach number, and contains uncertainties caused by perturbation of aerodynamic parameters, which enhances the coupling between channels and leads to more complex and difficult controller design.

3. Auto-Disturbance Rejection Control Technology Based on MIMO-ESO

3.1 MIMO-ESO ADRC design

For system (1), the design of the extended state observer ESO is the core of the entire auto-disturbance rejection controller. Each channel is a first-order system, so only the second-order extended state observer ESO needs to be reconstructed. The input matrix parameter B of the control variable δ is related to the aerodynamic parameter. Although there are related parameters that can be referenced, it is not an exact value, so B takes the reference aerodynamic parameter as the nominal value B_0 .Since $f(\boldsymbol{x})$ also has parameter perturbations and uncertainties, it is replaced by $f(\square)$. $f(\square)$ is the total uncertainty, including $f(\boldsymbol{x})$ and external environmental disturbances, unmodeled dynamics, coupling effects, and other disturbances.

Replace $g(x)$ with the estimated value g_0 , $f(\square)$ instead of $f(\boldsymbol{x})$, the system (1) can be equivalent to:

$$\dot{\boldsymbol{x}} = a(t) + U \tag{2}$$

Among them:

$$a(t) = f(\square) + (g(x) - g_0)\delta$$
$$U = g_0\delta$$

The system measurement equation is:

$$y = x = [\alpha \quad \beta \quad \gamma_c \quad \omega_x \quad \omega_y \quad \omega_z]^T$$

The second-order ESO with the same structure and the same parameters is configured for the six channels of the system (2). The generalized second-order MIMO-ESO equation is:

$$\begin{aligned} \mathbf{e}_1 &= \mathbf{z}_{11} - x \\ \dot{\mathbf{z}}_{11} &= \mathbf{z}_{12} - b_{11}\mathbf{e}_1 + U \\ \dot{\mathbf{z}}_{12} &= -b_{12}\mathbf{fal}(\mathbf{e}_1, a_1, \sigma_1) \end{aligned} \tag{3}$$

Among them $b_{11} > 0$, $b_{12} > 0$, $0 < a_1 < 1$, σ_1 are all scalar. $\mathbf{e}_1, \mathbf{z}_{11}, \dot{\mathbf{z}}_{11}, \mathbf{z}_{12}, \dot{\mathbf{z}}_{12}$ are all 6×1 vectors.

$$\mathbf{e}_1=\begin{bmatrix} e_{11}\\ e_{12}\\ e_{13}\\ e_{14}\\ e_{15}\\ e_{16}\end{bmatrix},\mathbf{fal}(\mathbf{e}_1,a_1,\sigma_1)=\begin{bmatrix}\mathrm{fal}(e_{11},a_1,\sigma_1)\\ \mathrm{fal}(e_{12},a_1,\sigma_1)\\ \mathrm{fal}(e_{13},a_1,\sigma_1)\\ \mathrm{fal}(e_{14},a_1,\sigma_1)\\ \mathrm{fal}(e_{15},a_1,\sigma_1)\\ \mathrm{fal}(e_{16},a_1,\sigma_1)\end{bmatrix}$$

The nonlinear function $\mathrm{fal}(e,a,\sigma)$ in the above formula is:

$$fal(e,a,\delta)=\begin{cases} e\times\delta^{a-1} & |e|\le\delta\\ |e|^a\,\mathrm{sgn}(e) & |e|>\delta\end{cases}$$

Among them $\mathrm{sgn}(e)$ is a symbolic function of e.

As long as the parameter selection in the ESO is appropriate and satisfies the following second-order ESO stability conditions: $b_{11}^2>4b_{12}\sigma_1^{a_1-1}$, then in the steady state, the ESO state will have the following convergence relationship:

$$\mathbf{z}_{11}\to x,\ \mathbf{z}_{12}\to a(t)=f(\square)+(g(x)-g_0)\delta$$

The ESO status will estimate the projectile angular velocity and the total uncertainties $a(t)$ in the model in real time. Using the estimated values $\mathbf{z}_{12}$, implement the following dynamic feedback compensation laws for system (2):

$$U=U_0-\mathbf{z}_{12} \tag{4}$$

The system (2) is then dynamically linearized into a single-integral system: $\dot{x}=U_0$.

It can be seen that after dynamic compensation, the six channels from control input U_0 to output x become parallel six single integrator systems, six channels are decoupled, and this dynamic feedback compensation is implemented using ESO without known $f(\square)$ and accurate parameter values. This method allows parametric perturbations, uncertainties, and interference effects, eliminating the need for accurate dynamic modelling of the body's attitude.

In order to quickly suppress the compensation residual, the control law uses a nonlinear state error feedback law (NLSEF) with a nonlinear (non-smooth) feedback effect to control the integral system.

$$\begin{aligned}\mathbf{e}_0&=x^*-\mathbf{z}_{11}\\ U_0&=b_0\mathbf{fal}(\mathbf{e}_0,a_0,\sigma_0)\end{aligned} \tag{5}$$

Among them, $\mathbf{e}_0=\begin{bmatrix}e_{01} & e_{02} & e_{03} & e_{04} & e_{05} & e_{06}\end{bmatrix}^T$, $x^*=[\alpha^*\ \ \beta^*\ \ \gamma_c^*\ \ \omega_x^*\ \ \omega_y^*\ \ \omega_z^*]^T$

In the above formula, $b_0>0$, $0<a_0<1$, σ_0 can be adjusted to avoid flutter near the zero point. $\left[\alpha^*,\beta^*,\gamma_c^*\right]$ is the desired attitude angle, and $\left[\omega_x^*,\omega_y^*,\omega_z^*\right]$ is the desired angular velocity needed to track the desired attitude angle, which is the real-time simulation quantity.

From equation (2), the control command for the vehicle's steering surface δ can be obtained (the following formula is called static decoupling law):

$$\delta=g_0^{-1}U \tag{6}$$

Equations (3) to (6) constitute an auto-disturbance-distorting attitude controller.

3.2 Stability Analysis and Error Analysis

System (1) becomes the following system under the action of the MIMO-ESO auto disturbance rejection controller:

$$\begin{aligned}\dot{x}&=a(t)+U_0-\mathbf{z}_{12}\\ U_0&=b_0\mathbf{fal}(\mathbf{e}_0,a_0,\sigma_0)\end{aligned} \tag{7}$$

Among them:

$$\mathbf{e}_1=\mathbf{z}_{11}-x,\ \mathbf{e}_2=\mathbf{z}_{12}-\mathbf{a(t)}$$

$$\dot{\mathbf{e}}_1=\mathbf{e}_2-b_{11}\mathbf{e}_1,\ \dot{\mathbf{e}}_2=-b_{12}\mathbf{fal}(\mathbf{e}_1,a_1,\sigma_1)-\dot{a}(\mathbf{t})$$

The following demonstrates the stability of the auto-disturbance rejection control system. First define the following function:

$$V_1=\frac{1}{2}\sum_{i=1}^{m}g_{2i}^2(e_{1i},e_{2i}) \tag{8}$$

$$g_{2i}(e_{1i},e_{2i})=\begin{cases}|h_{2i}(e_{1i},e_{2i})| & |h_{2i}(e_{1i},e_{2i})|>g_{1i}(e_{1i})\\ g_{1i}(e_{1i}) & |h_{2i}(e_{1i},e_{2i})|\le g_{1i}(e_{1i})\end{cases} \tag{9}$$

$$h_{2i}(e_{1i},e_{2i})=e_{2i}-b_{11}e_{1i}+kg_{1i}(e_{1i})\,\mathrm{sgn}(e_{1i}) \tag{10}$$

$$g_{1i}(e_{1i})=\frac{b_{12}}{kb_{11}}|fal(e_{1i},a_1,\sigma_1)|,\ k>1,\ i=1,\cdots,m \tag{11}$$

Lemma 1[13]: $G_2=\{(e_{1i},e_{2i}):|h_{2i}(e_{1i},e_{2i})|\le g_{1i}(e_{1i})\}$ is the self-stabilizing domain of the system.

Theorem 1: For the system (7), suppose that the expansion state quantity $a(t)$ is bounded by the time variation $-\dot{a}(t)=\left[-\dot{a}_1(t),\dots,-\dot{a}_m(t)\right]^T$, ie, if $|-\dot{a}_i(t)|<w$, $i=1,2,\cdots,m$ then $w>0$, if the system satisfies:

$$b_{11}^2>(1+k)c_2b_{12}\sigma^{a_1-1} \tag{12a}$$

$$b_{11}g_{2i}>\frac{c_2}{c_2-1}w\left(c_2>1,i=1,\cdots,m\right) \tag{12b}$$

Then $\dot{V}_1<0$, the system is a stable system.

Provement: $\dot{V}_1=\sum_{i=1}^{m}[g_{2i}(e_{1i},e_{2i})\dot{g}_{2i}(e_{1i},e_{2i})]$,

When $|h_{2i}(e_{1i},e_{2i})|>g_{1i}(e_{1i})$, then,

$$\begin{aligned}\dot{V}_{1i}&=g_{2i}(e_{1i},e_{2i})\dot{g}_{2i}(e_{1i},e_{2i})=h_{2i}\dot{h}_{2i}\\
&=h_{2i}\left[\frac{\partial h_{2i}}{\partial e_{2i}}\cdot\dot{e}_{2i}+\frac{\partial h_{2i}}{\partial e_{1i}}\cdot\dot{e}_{1i}\right]\\
&<h_{2i}w+\frac{\partial h_{2i}}{\partial e_{1i}}h_{2i}^2-h_{2i}\left(b_{12}\frac{|fal_{1i}|}{g_{1i}}+k\frac{\partial h_{2i}}{\partial e_{1i}}\right)g_{1i}\,\mathrm{sgn}(e_{1i})\\
&=h_{2i}w+\frac{\partial h_{2i}}{\partial e_{1i}}h_{2i}^2+k^2h_{2i}g_{1i}\left|\frac{dg_{1i}}{de_{1i}}\right|\\
&\le h_{2i}w-b_{11}h_{2i}^2+kh_{2i}^2\left|\frac{dg_{1i}}{de_{1i}}\right|+k^2h_{2i}g_{1i}\left|\frac{dg_{1i}}{de_{1i}}\right|\end{aligned}$$

Substituting equation (12a)、(12b) for availability:

$$\dot{V}_{1i} \le \frac{(c_2-1)}{c_2} b_{11}h_{2i}g_{1i} - b_{11}h_{2i}^2 + (1+k)h_{2i}^2 \frac{b_{12}}{b_{11}}\left|\frac{dfal_{1i}}{de_{1i}}\right|$$

$$= -\frac{1}{c_2} b_{11}h_{2i}^2 + (1+k)h_{2i}^2 \frac{b_{12}}{b_{11}}\left|\frac{dfal_{1i}}{de_{1i}}\right| < 0$$

Established by the above formula, can be set:

$$b_{11}^2 > (1+k)c_2 b_{12}\left|\frac{dfal_{1i}}{de_{1i}}\right| \quad (13)$$

If $dfal_{1i} = |e_{1i}|^{a_1}$,set $b_{11}^2 > (1+k)c_2 b_{12} a_1 |e_{1i}|^{a_1 - 1}$,

When $e_{1i} \to 0$,then $b_{11} \to \infty$.

To substitute $fal_{1i} = \begin{cases} |e_{1i}|^{a_1}\,\mathrm{sgn}(e_{1i}) & |e_{1i}| > \sigma \\ e_1\sigma^{a_1-1} & |e_{1i}| \le \sigma \end{cases}$ for this, you can get $b_{11}^2 > (1+k)c_2 b_{12}\sigma^{a_1-1}$, when formula (13) is satisfied, $\dot{V}_{1i} < 0$ and $\dot{V}_1 = \sum_1^m \dot{V}_{1i} < 0$.

When $|h_{2i}(e_{1i}, e_{2i})| \le g_{1i}(e_{1i})$, then,

$$|e_{2i} - b_{11}e_{1i} + kg_{1i}\mathrm{sgn}(e_{1i})| \le g_{1i}$$

$$-kg_{1i}\mathrm{sgn}(e_{1i}) - g_{1i} \le e_{2i} - b_{11}e_{1i} \le -kg_{1i}\mathrm{sgn}(e_{1i}) + g_{1i}$$

$$\dot{V}_{1i} = \left(\frac{b_{12}}{kb_{11}}\right)^2 |fal_{1i}|\left|\frac{dfal_{1i}}{de_{1i}}\right|\dot{e}_{1i}\mathrm{sgn}(e_{1i})$$

$$= \left(\frac{b_{12}}{kb_{11}}\right)^2 |fal_{1i}|\left|\frac{dfal_{1i}}{de_{1i}}\right|\frac{e_{1i}\dot{e}_{1i}}{|e_{1i}|}$$

Set $\left(\frac{b_{12}}{kb_{11}}\right)^2 |fal_{1i}|\left|\frac{dfal_{1i}}{de_{1i}}\right| = F$, then,

$$\dot{V}_{1i} = Fe_{1i}\dot{e}_{1i} = Fe_{1i}\left(e_{2i} - b_{11}e_{1i}\right) \le Fe_{1i}\left(-kg_{1i}\mathrm{sgn}(e_{1i}) + g_{1i}\right)$$

$$= F\left(-kg_{1i}|e_{1i}| + g_{1i}e_{1i}\right)$$

$$< -F(k-1)g_{1i}|e_{1i}| < 0$$

Therefore, Theorem 1 proves.

According to the theorems (1) and (12b), $g_{1i} = \frac{c_2}{(c_2-1)b_{11}} w$ is the boundary of the self-stability domain.

$$\begin{cases} |h_{2i}| = \frac{c_2}{(c_2-1)b_{11}} w & |h_{2i}| > g_{1i} \\ g_{1i} = \frac{c_2}{(c_2-1)b_{11}} w & |h_{2i}| \le g_{1i} \end{cases} \quad (14)$$

As

$$e_1^* = \max\{|e_{1i}|\} = \left[\frac{kc_2 w}{b_{12}(c_2-1)}\right]^{\frac{1}{a_1}}$$

$$e_2^* = \max\{|e_{2i}|\} = b_{11}e_1^* - \frac{(k-1)c_2 w}{b_{11}(c_2-1)}$$

In the formula, e_1^* and e_2^* are the maximum estimation errors of the response expansion observer to the state quantity $x(t)$ and the expansion state quantity $a(t)$, respectively.

When $b_{11} > \frac{kc_2 w}{(c_2-1)}$, a_1 is smaller, e_1^* and e_2^* are smaller too. Which means the smaller B is, the more accurate the estimation of the disturbance of the extended observer on the sum is, and the stronger the anti-interference ability is.

Redefine the function $V_2 = \frac{1}{2}\left(x^* - x\right)^T\left(x^* - x\right)$,

$$\dot{V}_2 = \frac{1}{2}\left(\dot{x}^* - \dot{x}\right)^T\left(x^* - x\right) = \left(\dot{x}^* - a(t) - U_0 + z_{12}\right)^T\left(x^* - x\right)$$

$$= -V_0^T\left(x^* - x\right) + \left(\dot{x}^* - a(t) + z_{12}\right)^T\left(x^* - x\right)$$

$$= -b_0\sum_{i=1}^m \left|\left(x^* - x\right)\right|^{1+a_0} + \left(\dot{x}^* - e_2\right)^T\left(x^* - x\right)$$

Thus, when the following inequality is satisfied:

$$b_0\left|\left(x^* - x\right)_i\right| > \left|\left[\dot{x}^* + e_2\right]_i\right|$$

$\dot{V}_2 < 0$ is Established.

Among them, $[A]_i$ is the ith element of A.

Therefore, the static tracking error of the self-interference control system is

$$\left|\left(x^* - x\right)_i\right| < \left|\frac{\left[\dot{x}^* + e_2\right]_i}{b_0}\right|^{\frac{1}{a_0}} \quad (15)$$

When $b_0 > \left|\left[\dot{x}^* + e_2\right]_i\right|$, the smaller a_0 is, the smaller the tracking error is.

Definition $V = V_1 + V_2$, when satisfying both (12) and (15), $\dot{V} = \dot{V}_1 + \dot{V}_2 < 0$ the self-interference control system is stable.

4.Simulation analysis

A certain type of BTT vehicle model is used as an example for simulation. α^* and γ_c^* are shown in the guidance command signals corresponding to the solid lines in Fig. 1 and Fig. 3, respectively. The coordination requirement of the BTT vehicle directly leads to $\beta^* = 0$.

Select the desired input $\left[\alpha^*, \beta^*, \gamma_c^*\right] = [5, 0, 5]$. The range of the perturbation of the aerodynamic parameters in the simulation is ±50%. Assess the performance of the control system in the case of a wide range of aerodynamic parameters.

Select design parameters: $b_{11} = 80$, $b_{12} = 0.001$, $a_1 = 0.5$, $\sigma_1 = 0.01$; $b_0 = 200$, $a_0 = 1$, $\sigma_0 = 0.01$.

The simulation results are shown in Figure 1~Figure 3. Figure 1 shows the attack angle response curve, Figure 2 shows the slip angle response curve, and Figure 3 shows the tilt angle response curve. Among them, the solid line is the tracking curve under the standard parameters, the broken line is the tracking curve when the aerodynamic parameter is perturbed by 50%, and the dotted line is the tracking curve when the aerodynamic parameter is perturbed by -50%.

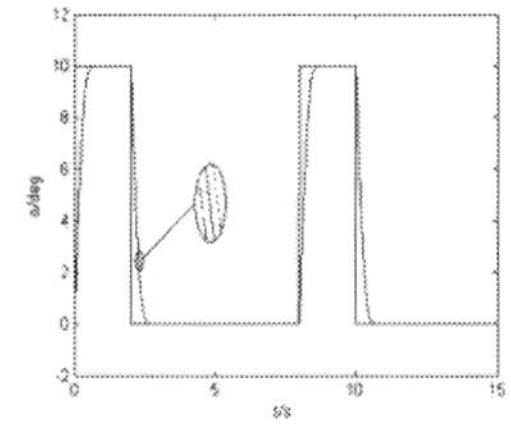

Figure 1. Attack angle curve

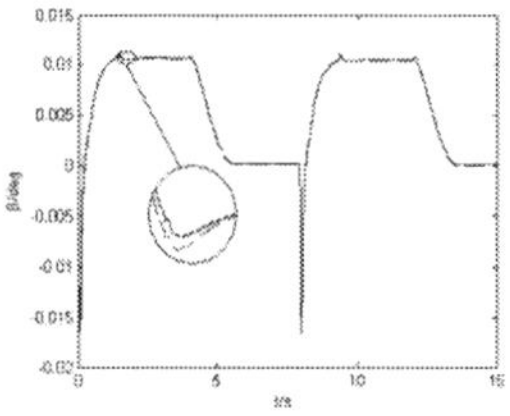

Figure 2. slip angle curve

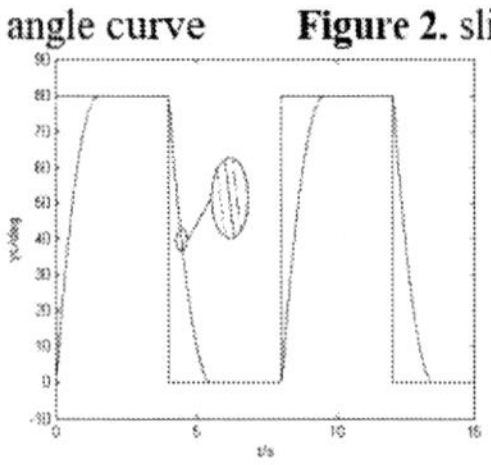

Figure 3. Tilt angle curve

The simulation results show that in the case of standard parameters, both the angle of attack and the angle of roll can track guidance command signal quickly and without overshooting, with good dynamic quality and high tracking accuracy, and the side slip angle satisfies the index $|\beta| \leq 0.3°$. The maximum rudder deflection angle of the vehicle is 10 degrees, and the maximum rudder deflection speed is 40 degrees/s. In the case of large perturbation of aerodynamic parameters, the three channels can also meet the stability requirements, and the control system still exhibits good stability and tracking performance, and has strong robustness.

5 Conclusion

In this paper, the MIMO-ESO auto disturbance-rejection attitude controller for hypersonic vehicle is designed based on the nonlinear attitude model for hypersonic vehicles' re-entry and the extended state observer and nonlinear state error feedback law in auto disturbance rejection control. Simulations show that the control system can overcome the influence of disturbance and aerodynamic parameters over a wide range of perturbations, and it has strong robustness while obtaining good dynamic quality and tracking performance. Because the extended state observer estimates the uncertainties, unmodeled dynamics, coupling effects, parametric perturbations, and interference effects as "summation interference" and dynamically compensates for feedback, the nonlinear residual state error feedback law is used to suppress the compensation residuals. The design of the ADRC is not required to accurately control the model of the vehicle, and only a standard value or an estimated value is required for the aerodynamic parameters. It is not necessary to know the perturbation limits of the aerodynamic parameters either. It successfully overcomes the difficulties in establishing an accurate controlled model and obtaining the parameter perturbation range in actual projects. Therefore, it has great engineering application value.

References

1. BOLENDER M A. An Overview on Dynamics and Controls Modelling of Hypersonic Vehicles. *2009 American Control Conference*,pp.2507-2512.
2. WANG Q,STENGEL R F.Robust nonlinear control of a hypersonic aircraft.*Journal of Guidance, Control and Dynamics, 2000,23*(4),pp.577–584.
3. ITO D,WARD D,VALASEK J. Robust dynamic inversion controller design and analysis for the X-38. *AIAA Guidance,Navigation,and Control Conference. Montreal, Canada,2001*.
4. XU H J,MIRMIRANI M,IOANNOU P A.Robust neural adaptive control of a hypersonic aircraft.*AIAA Guidance, Navigation ,and Control conference and Exhibit. Austin, Texas,2003*.
5. WALLNER E M,WELL K H.Nonlinear flight control design for the X-38 using CMAC neural networks.*AIAA Guidance, Navigation,and Control Conference and Exhibit. Montreal, Canada, 2001*.
6. XU H J,IOANNOU P A,MIRMIRANI M.Adaptive sliding mode control design for a hypersonic flight vehicle. *Journal of Guidance, Control and Dynamics, 2004,25*(5),pp.829–838.
7. Junchun Yang, Jun Hu, Xiaole Lv.Design of Sliding Mode Tracking Control for Hypersonic Reentry Vehicles. *Proceedings of the 26th Chinese Control Conference, Zhangjiajie, Hunan, China, 2007*.
8. Han J. From PID to active disturbance rejection control. *Industrial Electronics IEEE transactions on, 2009, 56*(3) ,pp.900-906.
9. Qi N M, Qin C M, Song Z G. Improved ADRC cascade decoupling controller design of hypersonic vehicle. *Harbin Gongye Daxue Xuebao(Journal of Harbin Institute of Technology), 2011, 43*(11),pp.34-38.
10. Herbst G. A Simulative Study on Active Disturbance Rejection Control (ADRC) as a Control Tool for Practitioners. *Electronics, 2013, 2*(3),pp.246-279.
11. Ti. TMS320F28069M. TMS320F28068M InstaSPIN(TM) MOTION Software Technical Reference Manual: *TEXAS INSTRUMENTS, 2013*.
12. Shaughnessy J D, Pinckney S Z, McMinn J D, Hypersonic vehicle simulation model: winged-cone configuration. *NASA TM-102610,1990*.
13. Huang Yi,Han Jing-qing.Analysis and design of nonlinear two-order continuous extended state observer.*Chinese Science Bulletin, 2000,45*(13), pp. 1373-1379.

Design and Control of a Quadruple-Input Double-Output Coherent Radar Test Platform Composed of Test & Measurement Equipment and Custom Hardware

Yanmin Chen, Kuoye Han and Mian Jiang

Information Science Academy(ISA), China Electronics Technology Group Corporation(CETC), Beijing 100098, P. R. China

Abstract. A radar test platform composed of commercial off-the-shelf test and measurement equipment and custom hardware has been built. This platform, which enables production of dual-channel coherent radar transmitting signals with a bandwidth up to 500MHz, and reception of four parallel channels of RF injection signals or echoes, can simulate various types of radar emitters or to simulate the multiple modes of operation for a single type radar. A software is also programmed with C++ language to control the waveforms, frequencies, phases, time delays and all other aspects of the test platform's operation at all stages. At last, an application example is given when the platform is used for ISAR imaging verification.

1 Introduction

Nowadays, the performance of telecommunication systems and radar systems has been greatly improved, due to the application of various new technologies, one of which is the technology of emitting and receiving multi-channel phase coherent signals. During the testing stage of the receivers of satellite communication systems, mobile communication systems, MIMO systems and phased array radar systems, it happened plenty of times that a set of multi-channel phase coherent signals is needed as a stimulus to the corresponding subsystems under test. These phase coherent signals are supposed to be with precisely adjustable frequencies, phases, and time delays, that can simulate the real-world signal characteristics and are used to test and evaluate the performance of the receivers or the subsystems. Therefore, it requires that a test platform capable of producing multi-channel phase coherent RF or IF signals flexibly and conveniently in the laboratory be built in order to implement the functions as mentioned above.

In this paper, we focus on the development and generation of actual radar test signals used to exercise all of the radar's functional stages either individually or the entire radar system collectively, for example, simulating the characteristics of a set of microwave signals at different input ports of array antennas, simulating the input signals of a digital beamforming network, or providing multi-channel coherent signal injections to the up- or down-conversion modules of a radar system and subsequently receiving and analysing the responses of the device under test. Signals can be applied at RF, IF or baseband to evaluate or characterize a radar or its subsystems.

A radar test platform composed of commercial off-the-shelf test and measurement equipment and custom hardware is designed and built. This platform supports the ability to generate dual-channel coherent transmitting signals and to receive four channels of test responses. By precisely controlling the waveforms, frequencies, phases, time delays and the signal flows, it can simulate various types of radar emitters or to simulate the multiple modes of operation for a single type of radar. The system design issues including the structure and functions, the composition of each functional module, the trigger and synchronization mechanism, the approach of maintaining phase consistency, and the control strategy and method are presented in the subsequent sections. At last, one of the I/O channels of the test platform is utilized, over which microwave pulses are sent and received, and an ISAR image of a maneuvering vehicle target is obtained, which collectively demonstrates the high quality of signals produced and good performance of the test platform.

2 Radar test platform design

2.1 System structure and functions

The radar test platform consists of the following modules: the dual-channel baseband signal generation module, the dual-channel quadrature (I/Q) modulation module, the frequency and timer module, the receiver down-conversion module, the AD sampling & recording module, the high power amplifier module and the antenna module, the power supply module and an industrial personal computer. Most core modules are made up of

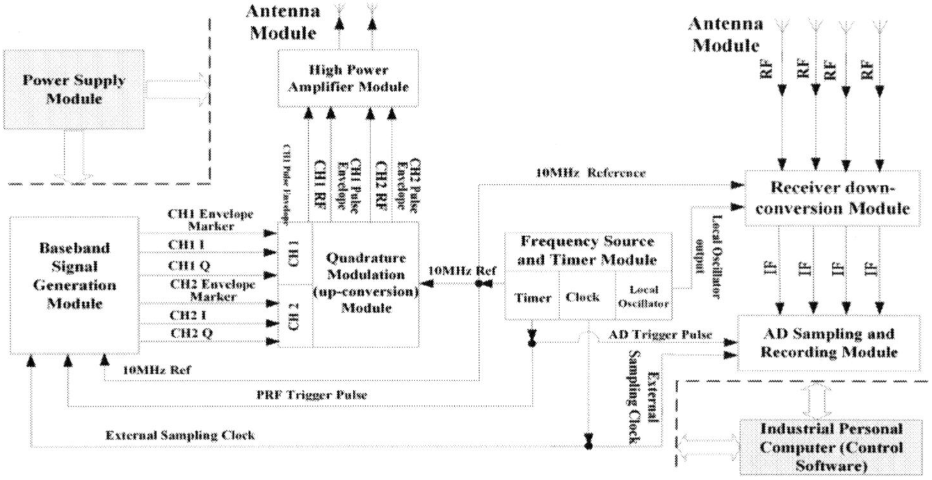

Figure 1. Block diagram of the radar test platform.

commercial off-the-shelf test & measurement equipment, which shows the broad capability of readily available hardware from a laboratory. Figure 1 is a block diagram of hardware modules in a configuration to support the ability to generate dual-channel coherent transmitting signals and to receive four channels of responses.

2.2 Baseband signal generation

The baseband signal generation module is based on arbitrary waveform generators (AWGs), where multiple set of I/Q waveform sequences can be stored in advance and then played as analogue signals with the actions of external triggers and clock. Flexible parameter setting gives the user choices of various types of waveforms including the monochromatic wave, linear frequency modulation (LFM) signal, and digital modulation signals frequently used in telecommunication such as BPSK, QPSK and etc.

As shown in Figure 2, two Agilent N6030A AWGs are configured to form a dual-channel baseband signal generation module. One N6030A unit is designated as the Master and the other one is designated as the Slave. These two AWGs are synchronized by sharing a common external clock and a common trigger, where either the trigger or the clock is split with low skew and distributed with cables of equal length to both units. The I or Q channel of each N6030A can produce signals range from DC to 500MHz, and the two quadrature signal components are synthesized to a signal with more broad bandwidth at the quadrature modulation module.

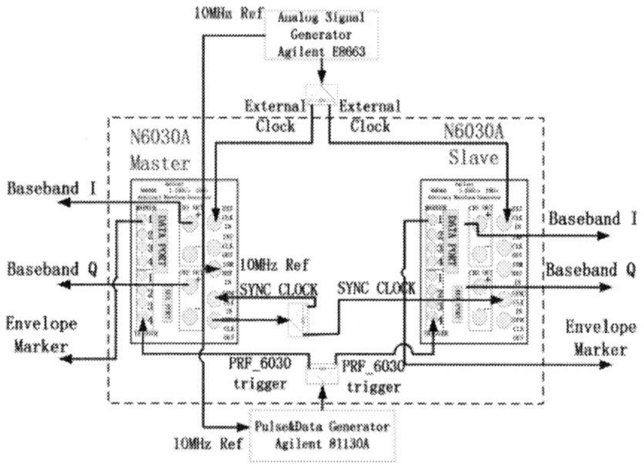

Figure 2. Configurations of the dual-channel baseband signal generation module.

2.3 Quadrature (I/Q) modulation: up-conversion

The quadrature (I/Q) modulation module is mainly made up of two Agilent E8267D PSG signal generators, one designated as Master and the other as Slave, corresponding to the two baseband signal generation channels, as shown in Figure 3.

The Master E8267D outputs a carrier frequency signal through an HCC option module installed on it, and this signal is split into two equivalents by a U3035P distribution network, then fed back to the input port of

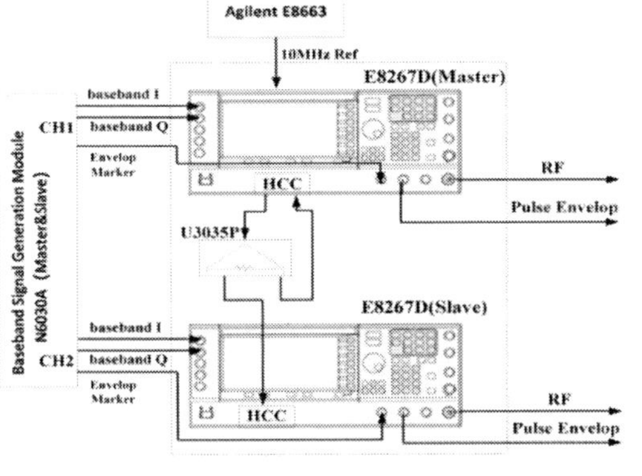

Figure 3. Configurations of the dual-channel baseband quadrature modulation module.

each E8267D's HCC option. By this means, the carrier frequency signals to the Master and the Slave channels maintain the same frequency and phase. It is also worth noting that an 015/016 wideband option is installed on the equipment to enable a production of signal bandwidth up to 1 GHz.

2.4 Receiver down-conversion

The receiver of the platform is designed to have four parallel channels, therefore, an Agilent N5280A four-channel frequency converter as shown in Figure 4 is chosen, with custom low noise RF amplifiers, band-pass filters and IF amplifiers integrated with it to down convert a set of received (or injected) RF signals.

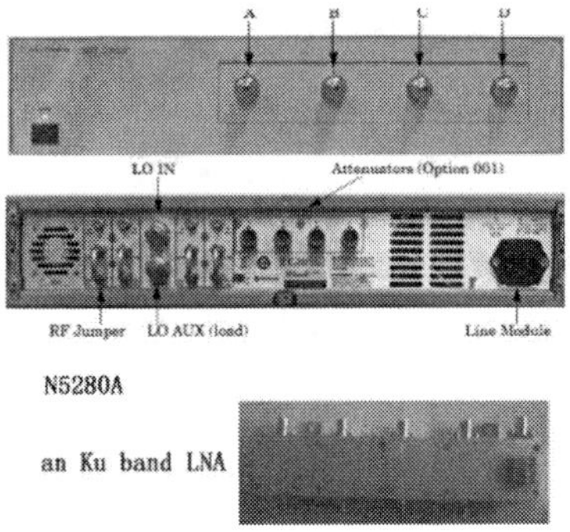

Figure 4. The N5280A four-channel frequency converter and a Ku band low noise amplifier.

2.5 Other modules

The frequency source and timer module is responsible for the generation of the triggers, the reference signals, the external clocks and the local oscillator. Here, an Agilent 81130A pulse generator is used to produce the triggers to the baseband signal generation module and to the AD sampling and recording module; an Agilent E8663 analog signal generator is used to produce the external clock signal, and the local oscillator to the receiver down-conversion module is produced by a R&S SMR20 analog signal generator. It is worth noting that the quadrature (I/Q) modulation module has its own carrier frequency source (E8268Ds), rather than using the output of the frequency source and timer module.

The high power amplifier module in this test platform consists of several Agilent 83020 high power amplifiers, which have a maximum gain of 30dB and can output signals with a maximum power of 30dBm.

The AD sampling and recording module is a four-channel custom module with a maximum sampling frequency of 2 GHz and maximum transmission bandwidth of 500MBps. It is designed to sample four IF down-converted IF signals simultaneously, then the sampled sequences are transmitted to the local industrial personal computer and digitally demodulated to I/Q channels.

The users can choose different antennas according to the specific applications, for example,

2.6 The synchronization mechanism and phase consistency

All the modules in the test platform share a common 10 MHz reference signal to guarantee the phase consistency among different local oscillators. Then each module puts this reference signal through its phase-lock-loop frequency multiplier or division circuit to generate its own signals. We have mentioned that the phase consistency among different parallel channels are held by using a Master-Slave form, for example, in the quadrature modulation module, the master and slave E8267Ds share a common carrier frequency originated from the master.

3 Access and control of the devices

Most of the commercial off-the-shelf test and measurement equipment use drives of the IVI-C or IVI-COM form and provide generalized Virtual Instrument Software Architecture (VISA) programming interface. The VISA is a Test & Measurement industry standard communication API (Application Programming Interface) for use with test and measurement devices. Figure 5 is a flow chart to initiate VISA and communicate with an equipment.

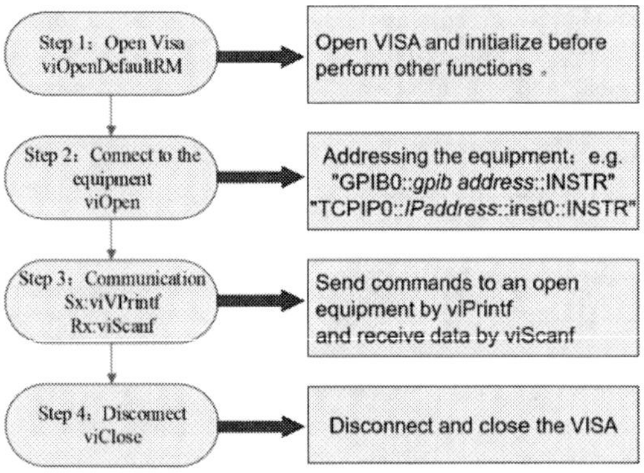

Figure 5. Control flow chart using VISA.

We developed a software, by which all the device of the test platform can be controlled individually and collectively. The software can adjust the signal parameters adaptively during the work period according to predefined patterns. It helps to simulate various types of radar emitters or to simulate the multiple modes of operation for a single type radar. The control software is programmed with C/C++ language and installed on an industrial personal computer. In addition, some signal/data processing functions are also integrated, e.g. digital quadrature demodulation, clutter filtering and pulse compression.

4 ISAR imaging verification experiment

An ISAR imaging verification experiment was carried out, where only one transceiver channel of the test platform is utilized. In Figure 6, the trigger pulse (yellow), the baseband I (green), baseband Q (blue), and the pulse envelop marker (pink) are displayed on an oscilloscope screen. Here, LFM I/Q baseband signal is generated from the AWG and transferred to the Quadrature (I/Q) modulation module for up-conversion on the carrier frequency. This transmission channel is pre-distortion calibrated to eliminate the I/Q imbalance and other channel imperfection. The platform works at Ku band and the measurement parameters are listed in Table 1.

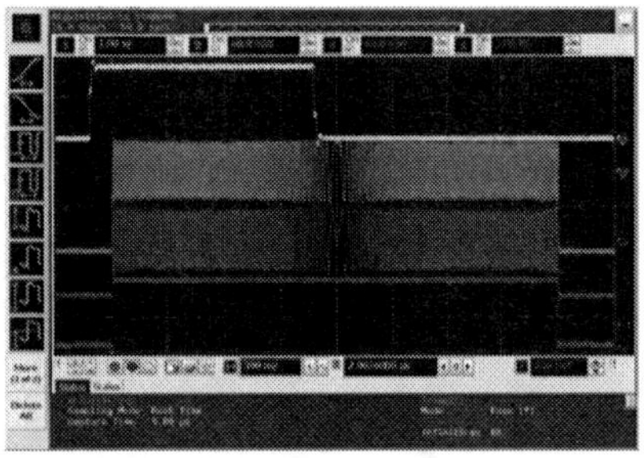

Figure 6. The trigger signal and signals generated by the baseband generation module.

Table 1. Measurement parameters.

Parameter	value	Parameter	value
Carrier frequency	17 GHz	Frequency bandwidth	400 MHz
PRF	6000 Hz	Peak power	0 dBm
LO frequency of down conversion	16.8 GHz	AD sampling frequency	960 MHz

We use two horn antennas, one for transmitting and the other for receiving, rather than using only one antenna and a circulator for transceiver channel switching. Before acquiring the ISAR echo data of a moving vehicle, we first turn the antenna beam at a corner reflector, see Figure 7, the range profile of this featured strong target is presented and the range resolution measured is close to the theoretical value of 0.375m. The response of this corner reflector can be saved as a reference for latter matched filtering processing. Also note that the first strong compressed pulse originates from direct coupling from the transmitting antenna to the receiving antenna, which should be supressed in practice.

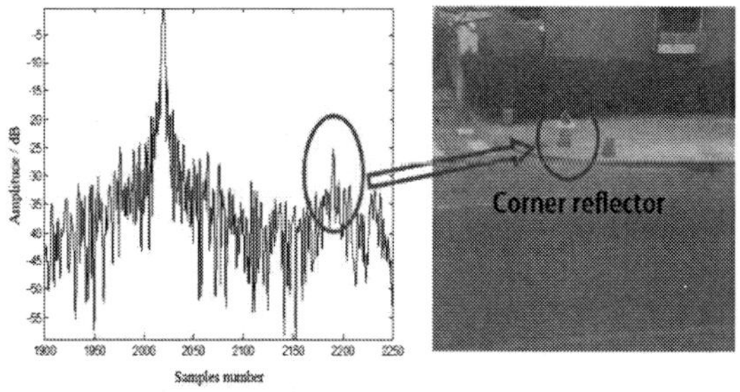

Figure 7. Range profile of a triangular corner reflector.

Figure 8. Target scenario: a moving vehicle.

Then ISAR echo data is acquired using the parameters listed in Table 1. The imaging process is performed in virtue of a motion compensation algorithm, by which each return range line is first aligned, then phase compensation based on the phase gradient autofocus (PGA) method is performed on each return. The imaging result of a moving vehicle on the scene is presented in Fig.9, the vehicle target was well focused and the outline is very clear, this indirectly demonstrates the good channel performance and phase consistency of the test platform.

https://doi.org/10.1051/matecconf/201821403004

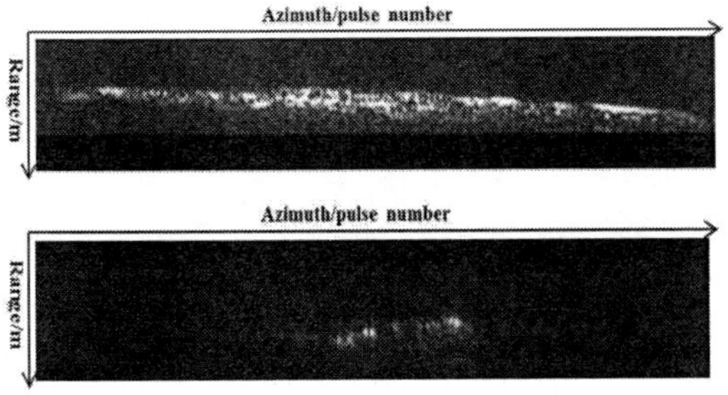

Figure 1. Envelope alignment and ISAR imaging result for a non-cooperative moving vehicle

5 Conclusion

A radar test platform is introduced. The platform is built upon commercial off-the-shelf test & measurement equipment and custom hardware. It can test the performance of radar hardware modules by providing injection signals or simulate a radar transceiver. A control software is also provided to control the platform adaptively. At last, real experiment result is given to demonstrate the good performance of this platform.

References

1. Barton, D.K., *Radar System Analysis* (Dedham, MA: Bartech House, 1979)
2. Agilent Technologies, *Agilent N6030A Series Arbitrary Waveform Generators User's Guide.* (2006)
3. Agilent *RF Signal Generator IVI-COM Driver Reference, Agilent Technologies*, (2006)
4. Su F. L., Cao Z. D.. *Improvements on the range alignment of the motion compensation for ISAR imaging*, IEEE International Radar Conference. New York: 707-710 (1996)

RBF NN-Based Backstepping Adaptive Control for a Class of Nonlinear Systems

Haibo Zhao[1], Chengguang Wang[2]

[1]*Engineering Technology Research Center of Optoelectronic Appliance, Tongling University, Tongling Anhui, 244061, China*
[2]*Sichuan Institute of Aerospace System Engineering, Chengdu Sichuan, 610100, China*

Abstract. There are many control methods for nonlinear systems, but some of them can not control nonlinear mismatched systems very well. Backstepping control has obvious advantages in controlling nonlinear mismatched systems. Thus we proposed a new radial-basis-function (RBF) neural network-based backstepping adaptive controller combining RBF neural network (RBF NN) and backstepping control for a class of nonlinear mismatched systems. We adopted RBF NN to approximate the system uncertainty. And we analyzed the controller stability using Lyapunov stability theory. Finally we chose sine signal as simulation input signal, simulation results show that the proposed control strategy has better adaptive ability and robustness than PID control, validating the effectivess of the proposed control strategy.

1 Introduction

Nonlinear systems have been studied by many scholars, and the control methods used are as follows: iterative learing control [1], feedback linearization [2], fuzzy control [3], sliding mode control [4], etc. However, when the system is a nonlinear mismatched system, the disadvantages of the above methods will be revealed.

Backstepping control can be used to control nonlinear mismatched systems [5]. RBF neural network (RBF NN) has universal approximation capability and can approximate arbitrary nonlinear functions with arbitrary precision [6]. The control algorithm combining RBF NN and backstepping control has been successfully applied by scholars [7-11].

In this paper, we use RBF NN-based backstepping control for a class of nonlinear mismatched systems. RBF NN is used to approximate system uncertainty. Simulation results verify the feasibility of the proposed control algorithm.

2 Problem description

Consider the following nonlinear mismatched system with strict feedback [12]:

$$\begin{cases} \dot{x}_1 = x_2 + f_1(x_1) + \Delta f_1(x_1) \\ \dot{x}_2 = x_3 + f_2(x_1,x_2) + \Delta f_2(x_1,x_2) \\ \cdots \\ \dot{x}_{n-1} = x_n + f_{n-1}(x_1,x_2,\cdots,x_{n-1}) + \Delta f_{n-1}(x_1,x_2,\cdots,x_{n-1}) \\ \dot{x}_n = \psi u + f_n(x_1,x_2,\cdots,x_n) + \Delta f_n(x_1,x_2,\cdots,x_n) \\ y = x_1 \end{cases} \tag{1}$$

where $\boldsymbol{x} = [x_1,x_2,\cdots,x_n]^{\mathrm{T}}$ is a system state variable. y is a system output. u is a system control input. Let $\bar{\boldsymbol{x}}_i = [x_1,x_2,\cdots,x_i]^{\mathrm{T}}$, strict feedback means that the nonlinear terms f_i and Δf_i depend only on the system state $x_1,x_2,\cdots,x_i$.

In addition, the nonlinear terms satisfy the following characteristics : (a) $f_i : R^i \to R$ should be smooth enough and $f_i(0,0,\cdots,0) = 0$, $i = 1,2,\cdots,n$; (b) $\Delta f_i : R^i \to R$ is unknown, $\Delta f_i \in C^l$ and $\Delta f_i(0,0,\cdots,0) = 0$.

Assumption 1 Given a smooth nonlinear function vector $\boldsymbol{h} : \Omega \mapsto R^n$ (Ω is a compact set in R^m), there exist an optimal Gaussian base function vector $\boldsymbol{\varphi} : R^m \mapsto R^n$ and weight matrix $\boldsymbol{W}^*_{n\times n}$ such that :

$$\boldsymbol{h}(\boldsymbol{x}) = \boldsymbol{W}^{*\mathrm{T}}\boldsymbol{\varphi}(\boldsymbol{x}) + \boldsymbol{\varepsilon}, \forall \boldsymbol{x} \in \Omega \tag{2}$$

where ε is the construction error of RBF NN.

In this paper, our main task is to design an adaptive controller so as to make system output x_1 asymptotically stable tracking the expected output x_{1d} , namely $\lim_{t\to\infty}|x_1 - x_{1d}| = 0$.

3 Controller design and stability analysis

We use a RBF NN-based backstepping adaptive control strategy for a class of nonlinear system, Lyapunov stability theory is used to analyze the system stability. The detailed design process is as follows.

Step 1 Define the tracking error between the actual system output x_1 and the expected system output x_{1d} as :

$$e_1 = x_1 - x_{1d} \tag{3}$$

Its derivative is

$$\dot{e}_1 = \dot{x}_1 - \dot{x}_{1d} = x_2 + f_1(x_1) + \Delta f_1(x_1) - \dot{x}_{1d} \quad (4)$$

To eliminate the influence of uncertainty Δf_1, we use RBF NN to approximate Δf_1 :

$$\Delta f_1(x_1) = \boldsymbol{W}_1^{*\mathrm{T}} \boldsymbol{\varphi}_1(x_1) + \varepsilon_1 \quad (5)$$

where x_1 is a RBF NN input.

If x_2 is considered the actual input of system (4), then there exist an ideal control x_2^* [13]:

$$x_2^* = -(f_1 - \dot{x}_{1d} + k_1 e_1 + \boldsymbol{W}_1^{*\mathrm{T}} \boldsymbol{\varphi}_1 + \varepsilon_1) \quad (6)$$

such that $\dot{e}_1 = -k_1 e_1 + (x_2 - x_2^*)$, $k_1 > 0$ is a design parameter. Because x_2^* is not available, we design a virtual control x_{2d} as follows.

$$x_{2d} = -(f_1 - \dot{x}_{1d} + k_1 e_1 + \hat{\boldsymbol{W}}_1^{\mathrm{T}} \boldsymbol{\varphi}_1) \quad (7)$$

where $\hat{\boldsymbol{W}}_1$ is the estimated value of $\boldsymbol{W}_1^*$.

We choose the following Lyapunov function V_1 :

$$V_1 = \frac{1}{2} e_1^2 + \frac{1}{2} tr[\tilde{\boldsymbol{W}}_1^{\mathrm{T}} \boldsymbol{\Gamma}^{-1} \tilde{\boldsymbol{W}}_1] \quad (8)$$

where $\tilde{\boldsymbol{W}}_1 = \boldsymbol{W}_1^* - \hat{\boldsymbol{W}}_1$, $\boldsymbol{\Gamma}_1 = \boldsymbol{\Gamma}_1^{\mathrm{T}} > 0$ is a design parameter.

The derivative of V_1 is as follows :

$$\dot{V}_1 = e_1 \dot{e}_1 + tr[\tilde{\boldsymbol{W}}_1^T \boldsymbol{\Gamma}_1^{-1} \dot{\tilde{\boldsymbol{W}}}_1]$$

$$= -k_1 e_1^2 + e_1(x_2 - x_{2d}) + \tilde{\boldsymbol{W}}_1^{\mathrm{T}} \boldsymbol{\varphi}_1 e_1 + \varepsilon_1 e_1 + tr[\tilde{\boldsymbol{W}}_1^{\mathrm{T}} \boldsymbol{\Gamma}_1^{-1} \dot{\tilde{\boldsymbol{W}}}_1] \quad (9)$$

Step *i* ($i = 2, \cdots, n-1$) Define the tracking error e_i between the system state x_i and the virtual control x_{id} as

$$e_i = x_i - x_{id} \quad (10)$$

Its derivative is

$$\dot{e}_i = \dot{x}_i - \dot{x}_{id} = x_{i+1} + f_i(\bar{x}_i) + \Delta f_i(\bar{x}_i) - \dot{x}_{id} \quad (11)$$

Similarly, to eliminate the influence of uncertainty Δf_i, we use RBF NN to approximate Δf_i :

$$\Delta f_i(\bar{x}_i) = \boldsymbol{W}_i^{*\mathrm{T}} \boldsymbol{\varphi}_i(\bar{x}_i) + \varepsilon_i \quad (12)$$

where $\bar{x}_i$ is a RBF NN input.

If x_{i+1} is considered the actual input of system (11), then there exist an ideal control x_{i+1}^* :

$$x_{i+1}^* = -(f_i - \dot{x}_{id} + k_i e_i + e_{i-1} + \boldsymbol{W}_i^{*\mathrm{T}} \boldsymbol{\varphi}_i + \varepsilon_i) \quad (13)$$

such that $\dot{e}_i = -k_i e_i - e_{i-1} + (x_{i+1} - x_{i+1}^*)$, $k_i > 0$ is a design parameter. Because x_{i+1}^* is not available, we design a virtual control $x_{i+1,d}$ as follows.

$$x_{i+1,d} = -(f_i - \dot{x}_{id} + k_i e_i + e_{i-1} + \hat{\boldsymbol{W}}_i^{\mathrm{T}} \boldsymbol{\varphi}_i) \quad (14)$$

where $\hat{\boldsymbol{W}}_i$ is the estimated value of $\boldsymbol{W}_i^*$.

We choose the following Lyapunov function V_i :

$$V_i = \sum_{j=1}^{i-1} V_j + \frac{1}{2} e_i^2 + \frac{1}{2} tr[\tilde{\boldsymbol{W}}_i^{\mathrm{T}} \boldsymbol{\Gamma}_i^{-1} \tilde{\boldsymbol{W}}_i] \quad (15)$$

where $\tilde{\boldsymbol{W}}_i = \boldsymbol{W}_i^* - \hat{\boldsymbol{W}}_i$, $\boldsymbol{\Gamma}_i = \boldsymbol{\Gamma}_i^{\mathrm{T}} > 0$ is a design parameter.

The derivative of V_i is as follows :

$$\dot{V}_i = -\sum_{j=1}^{i} k_j e_j^2 + e_i(x_{i+1} - x_{i+1,d})$$

$$+ \tilde{\boldsymbol{W}}_i^{\mathrm{T}} \boldsymbol{\varphi}_i e_i + \varepsilon_i e_i + tr[\tilde{\boldsymbol{W}}_i^{\mathrm{T}} \boldsymbol{\Gamma}_i^{-1} \dot{\tilde{\boldsymbol{W}}}_i] \quad (16)$$

Step *n* Define the tracking error e_n between the system state x_n and the virtual control x_{nd} as

$$e_n = x_n - x_{nd} \quad (17)$$

Its derivative is

$$\dot{e}_n = \dot{x}_n - \dot{x}_{nd} = \psi u + f_n(\bar{x}_n) + \Delta f_n(\bar{x}_n) - \dot{x}_{nd} \quad (18)$$

To eliminate the influence of uncertainty Δf_n, we use RBF NN to approximate Δf_n :

$$\Delta f_n(\bar{x}_n) = \boldsymbol{W}_n^{*\mathrm{T}} \boldsymbol{\varphi}_n(\bar{x}_n) + \varepsilon_n \quad (19)$$

u is the actual control of system (18), there exist an ideal control u^* :

$$u^* = -(f_n - \dot{x}_{nd} + k_n e_n + e_{n-1} + \boldsymbol{W}_n^{*\mathrm{T}} \boldsymbol{\varphi}_n + \varepsilon_n) \quad (20)$$

such that $\dot{e}_n = -k_n e_n - e_{n-1} + \psi(u - u^*)$, $k_n > 0$ is a design parameter. Because u^* is not available, we design an actual control u as follows.

$$u = -(f_n - \dot{x}_{nd} + k_n e_n + e_{n-1} + \hat{\boldsymbol{W}}_n^{\mathrm{T}} \boldsymbol{\varphi}_n) \quad (21)$$

where $\hat{\boldsymbol{W}}_n$ is the estimated value of $\boldsymbol{W}_n^*$.

We choose the following Lyapunov function V_n :

$$V_n = \frac{1}{2} \boldsymbol{\xi}^{\mathrm{T}} \boldsymbol{\xi} + \frac{1}{2} tr[\tilde{\boldsymbol{Z}}^{\mathrm{T}} \boldsymbol{\Gamma}^{-1} \tilde{\boldsymbol{Z}}] + \frac{1}{2} \eta \tilde{\psi}^2 \quad (22)$$

where $\eta > 0$ is a design parameter, $\boldsymbol{\Gamma} = \begin{bmatrix} \boldsymbol{\Gamma}_1 & & & \\ & \boldsymbol{\Gamma}_2 & & \\ & & \ddots & \\ & & & \boldsymbol{\Gamma}_n \end{bmatrix}$,

$\boldsymbol{\xi} = [e_1\ e_2 \cdots e_n]^{\mathrm{T}}$, and $\tilde{\psi} = \psi - \hat{\psi}$, $\hat{\psi}$ is the estimated value of ψ .

We choose the following adaptive law :

$$\dot{\hat{\boldsymbol{Z}}} = \boldsymbol{\Gamma} \boldsymbol{\varphi} \boldsymbol{\xi}^{\mathrm{T}} - n \boldsymbol{\Gamma} \|\boldsymbol{\xi}\| \hat{\boldsymbol{Z}} \quad (23)$$

where $\boldsymbol{\varphi} = [\boldsymbol{\varphi}_1\ \boldsymbol{\varphi}_2 \cdots \boldsymbol{\varphi}_n]$, $n > 0$ is a design parameter.

$$\boldsymbol{Z} = \begin{bmatrix} \boldsymbol{W}_1 & & & \\ & \boldsymbol{W}_2 & & \\ & & \ddots & \\ & & & \boldsymbol{W}_n \end{bmatrix}, \ \|\boldsymbol{Z}\|_F \le Z_M$$

$$\hat{\boldsymbol{Z}} = \begin{bmatrix} \hat{\boldsymbol{W}}_1 & & & \\ & \hat{\boldsymbol{W}}_2 & & \\ & & \ddots & \\ & & & \hat{\boldsymbol{W}}_n \end{bmatrix}, \ \tilde{\boldsymbol{Z}} = \boldsymbol{Z} - \hat{\boldsymbol{Z}}$$

The derivative of V_n is as follows :

$$\dot{V}_n = -\boldsymbol{\xi}^{\mathrm{T}} \boldsymbol{K}_e \boldsymbol{\xi} + \boldsymbol{\xi}^{\mathrm{T}} \boldsymbol{\varepsilon} + tr[\tilde{\boldsymbol{Z}}^{\mathrm{T}} \boldsymbol{\Gamma}^{-1} \dot{\tilde{\boldsymbol{Z}}} + \tilde{\boldsymbol{Z}}^{\mathrm{T}} \boldsymbol{\varphi} \boldsymbol{\xi}^{\mathrm{T}}] + \eta \tilde{\psi} \dot{\tilde{\psi}} \quad (24)$$

where

$\boldsymbol{K}_e = diag\{k_1, k_2, \cdots, k_n\}^{\mathrm{T}}$, $\boldsymbol{\varepsilon} = [\varepsilon_1\ \varepsilon_2 \cdots \varepsilon_n]^{\mathrm{T}}$, $\|\boldsymbol{\varepsilon}\| < \varepsilon_N$.

Since $\dot{\tilde{\boldsymbol{Z}}} = -\dot{\hat{\boldsymbol{Z}}}$, and $\dot{\tilde{\psi}} = -\dot{\hat{\psi}}$, substitute Eq.(23) into (24), we have

$$\dot{V}_n = -\boldsymbol{\xi}^{\mathrm{T}} \boldsymbol{K}_e \boldsymbol{\xi} + \boldsymbol{\xi}^{\mathrm{T}} \boldsymbol{\varepsilon} + n\|\boldsymbol{\xi}\| tr[\tilde{\boldsymbol{Z}}^{\mathrm{T}} (\boldsymbol{Z} - \tilde{\boldsymbol{Z}})] - \eta \tilde{\psi} \dot{\hat{\psi}} \quad (25)$$

According to the actual system, the inequality $\tilde{\psi} \dot{\hat{\psi}} \ge 0$ is clearly true. In addition, according to Schwarz inequality, the following inequality is true :

$$tr[\tilde{\boldsymbol{Z}}^{\mathrm{T}} (\boldsymbol{Z} - \tilde{\boldsymbol{Z}})] \le \|\tilde{\boldsymbol{Z}}\|_F \|\boldsymbol{Z}\|_F - \|\tilde{\boldsymbol{Z}}\|_F^2$$

Since $K_{\min}\|\xi\|^2 \le \xi^T \boldsymbol{K}_e \xi$, where $K_{\min}$ is the minimum eigenvalue of $\boldsymbol{K}_e$. Thus the above Eq.(25) becomes :

$$\dot{V}_n \le -K_{\min}\|\xi\|^2 + \varepsilon_N\|\xi\| + n\|\xi\|(\|\tilde{\boldsymbol{Z}}\|_F\|\boldsymbol{Z}\|_F - \|\tilde{\boldsymbol{Z}}\|_F^2) - \eta\tilde{\psi}\dot{\hat{\psi}}$$
$$\le -\|\xi\|(K_{\min}\|\xi\| - \varepsilon_N + n\|\tilde{\boldsymbol{Z}}\|_F(\|\tilde{\boldsymbol{Z}}\|_F - Z_M)) \tag{26}$$

As we know,

$$K_{\min}\|\xi\| - \varepsilon_N + n(\|\tilde{\boldsymbol{Z}}\|_F^2 - \|\tilde{\boldsymbol{Z}}\|_F Z_M)$$
$$= K_{\min}\|\xi\| - \varepsilon_N + n(\|\tilde{\boldsymbol{Z}}\|_F - \frac{1}{2}Z_M)^2 - \frac{n}{4}Z_M^2$$

Thus if the following condition can be met

$$\|\xi\| > \frac{\varepsilon_N + \frac{n}{4}Z_M^2}{K_{\min}} \tag{27}$$

then $\dot{V}_n \le 0$.

From Eq.(27), we can see that the tracking accuracy is related to ε_N, n and $K_{\min}$.

4 Simulation analysis

We adopt the following system to validate the effectiveness of the proposed control strategy.

$$\begin{cases} \dot{x}_1 = x_2 + x_1\sin(x_1) \\ \dot{x}_2 = x_3 + x_1^2 + x_2^2 + \sin(x_1 + x_2) \\ \dot{x}_3 = 2u + \sin(x_1^2 + x_2^2 + x_3^2) \\ y = x_1 \end{cases} \tag{28}$$

The control goal is to make the system output x_1 trace the desired output $x_{1d} = 0.5\sin(2\pi t)$. $k_1 = 10$, $k_2 = 8$, $k_3 = 12$, $\Gamma_i(i = 1,2,3)$ is a unit array with proper dimension, $\eta = 115$, $n = 0.01$, $\boldsymbol{x}(0) = [0,0,0]^T$. The following curves are comparision curves between PID control and the proposed strategy in this paper, namely RBF NN-based backstepping adaptive control(RBFNNB). Figure 1 and figure 2 are curves of tracking sine signal and error using PID control, respectively. Figure 3 and figure 4 are curves of tracking sine signal and error using RBFNNB, respectively. From figure 1-4, it is obvious for us that RBFNNB has better dynamic and steady performance than PID control.

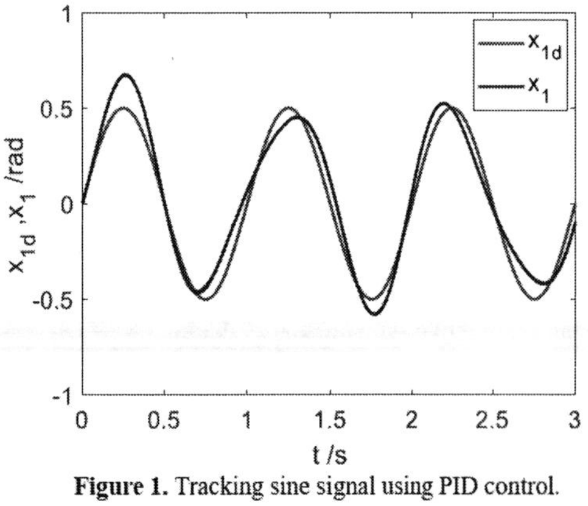

Figure 1. Tracking sine signal using PID control.

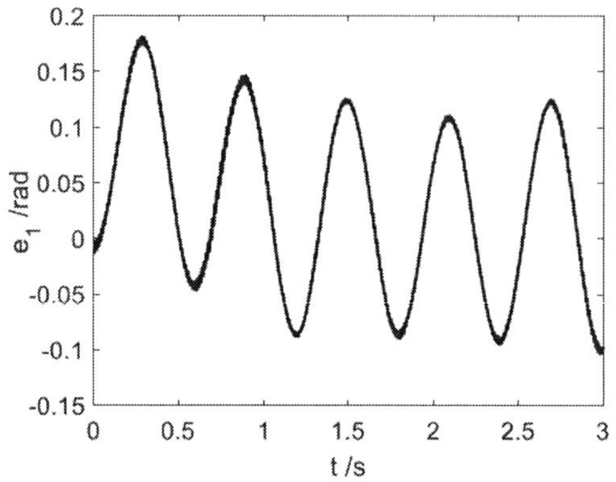

Figure 2. Error of tracking sine signal using PID control.

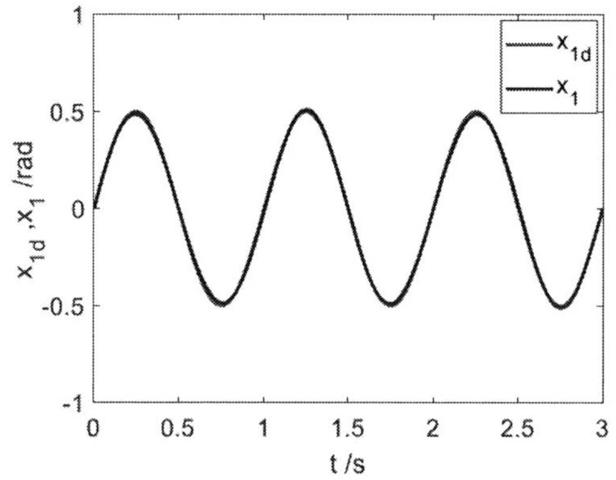

Figure 3. Tracking sine signal using RBFNNB.

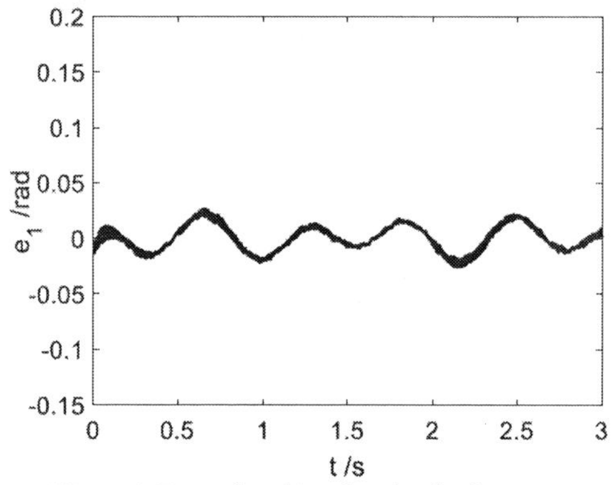

Figure 4. Error of tracking sine signal using RBFNNB.

Conclusion

For a class of nonlinear mismatched systems, we proposed a new RBF NN-based backstepping adaptive control. We designed the adaptive controller combining RBF NN and backstepping control, and analyzed its stability. In simulation analysis, sine signal is chosen as input signal, system has larger tracking error and worse adaptive tracking ability when using PID control, but system has smaller tracking error and better adaptive tracking ability when using RBFNNB, validating the superiority of the proposed control strategy. However, the system studied in this paper is the single input and single output (SISO) system, there is also multiple input and multiple output (MIMO) system in practical engineering. In this paper, we do not consider MIMO system, and this will be the focus of our future work.

Conflict of interest

The authors confirm that this article content has no conflict of interest.

Acknowledgements

This work was supported by National Natural Science Foundation of China(No.61074023), Natural Science Research Key Project of Universities in Anhui Province of China (No.KJ2015A297), Nature Science Foundation in Anhui Province of China (No.1508085MF130), the Engineering Technology Research Center of Optoelectronic Appliance in Anhui Province of China, Sichuan Institute of Aerospace System Engineering in Sichuan Province of China.

References

1. J. Wei, Y. Zhang, H. Bao, An exploration on adaptive iterative learning control for a class of commensurate high-order uncertain nonlinear fractional order systems, IEEE CAA J. Autom. Sin., 5(2) (2018) 618-627.
2. T. Kaczorek, Positivity and linearization of a class of nonlinear continuous-time systems by state feedbacks, Int. J. Appl. Math. Comput. Sci., 25(4) (2015) 827-831.
3. W. Lv, F. Wang, Finite-time adaptive fuzzy tracking control for a class of nonlinear systems with unknown hysteresis, Int. J. Fuzzy Syst., 20(3) (2018) 782-790.
4. J. Zhang, M. Lyu, T. Shen, Sliding mode control for a class of nonlinear multi-agent system with time delay and uncertainties, IEEE Trans Ind Electron, 65(1) (2018) 865-875.
5. K. Zhu, N.M. Qi, C.M. Qin, Adaptive sliding mode controller design for BTT missile based on backstepping control, Yuhang Xuebao, 31(3) (2010) 769-773.
6. J. Park, I.W. Sandberg, Universal approximation using radial-basis-function networks, Neural Computation, 3(2) (1991): 246-257.
7. S. Zhang, Z. Wu, An adaptive block backstepping control strategy for motorized spindles based on improved RBF neural network, Boletin Tecnico, 55(9) (2017) 432-439.
8. J. Liu, Radial basis function (RBF) neural network control for mechanical systems: Design, analysis and matlab simulation[M], Germany: Springer-Verlag Berlin Heidelberg, 2013.
9. W. Si, X. Dong, F. Yang, Decentralized adaptive neural control for high-order stochastic nonlinear strongly interconnected systems with unknown system dynamics, Inf Sci, 424 (2018) 137-158.
10. Y. Sun, B. Chen, C. Lin, Adaptive neural control for a class of stochastic nonlinear systems by backstepping approach, Inf Sci , 369 (2016) 748-764.
11. D. Mu, G. Wang, Y. Fan, Design of robust adaptive course controller for unmanned surface vehicle with input saturation, Int. J. Innov. Comput. Inf. Control, 13(5) (2017) 1751-1758.
12. X.J. Ren, R.C. Liu, Y. Chen, Z.Y. Yang, Robust adaptive control for a class of nonlinear systems using backstepping based on RBF neural network, Journal of Naval Aeronautical and Astronautical University, 23(6) (2008) 645-648.
13. Y. Hu, Y. Jin, P. Cui, RBF NN-based backstepping control for strict feedback block nonlinear system and its application, Lect. Notes Comput. Sci., 3174 (2004) 129-137.

Initial Alignment of Large Azimuth Misalignment Angle in SINS based on Reduced Multiple Fading Factors Strong Tracking CKF

LU Hang, HAO Shun-yi, PENG Zhi-ying and HUANG Guo-rong

College of Aeronautics and Astronautics Engineering, Air Force Engineering University, Shanxi, Xi'an 710038, China

Abstract. Aiming at the problem that Cubature Kalman Filter(CKF) has low accuracy and robustness under the condition of strapdown inertial navigation system(SINS) initial alignment due to model error and external disturbance, Reduced Multiple Strong Tracking Cubature Kalman Filter(RMSTCKF) is proposed, and the algorithm flow and sub-optimal solution of multiple fading factor are derived. Multiple fading factor can improve tracking ability under each state according to the degree of uncertainty of different states, having stronger adaptability and robustness. Applying RMSTCKF to large azimuth misalignment angle error function described by Euler platform error angle(EPEA), carrying out the simulation under two different conditions, namely noise mismatch and the base is disturbed, and making contrast between RSTCKF and RCKF, the simulation results show that the filter accuracy and convergence rate of RMSTCK when system noise mismatches with true noise are obviously better than RSTCKF and RCKF, having better practical value in engineering.

1 INTRODUCTION

Practical working environment of SINS is usually complex and uncertain, the result of SINS coarse alignment is large misalignment, thus making linear error model and KF algorithm inapplicable based on the assumption of small misalignment. Article [1] uses the nonlinear error model and nonlinear filter algorithms described by EPEA to solve the problem, but system will be effected by model error, noise and external disturbance, so nonlinear filter sill have low robustness and tracking capability[2,3]. Article [4] proposes strong tracking filter theory, STF can maintain stronger robustness and tracking capability even when facing model uncertainty, noise and big external disturbance[3,4,5]. Article [3] applies STFUKF to initial alignment, but uncertainty of each state vector is different, so it proves that single fading factor can't track each state vector perfectly. MSTUKF is proposed in article [7], which doesn't need to calculate complex Jacobi matrix. Current theory has proved that the accuracy of UKF and CKF can reach the second-order and the third-order respectively, meanwhile, CKF needn't more sampling points and adjustable parameters, having shorter filter time and higher numerical stability[8,9,10].

This article derives reduced CKF on the basis that the distribution characteristics of cubature points in the third-order spherical radial cubature principle and the measurement equation is linear, combing STF with RCKF framework, introducing 2 multiple fading factor matrix, and proposing RMSTCKF that is appropriate to initial alignment. RMSTCKF only needs one cubature transformation without calculating Jacobi matrix, and multiple fading factor can improve tracking ability correspondingly according to the degree of uncertainty in different states. The SINS nonlinear error function described by EPEA in article [1] is applied in this article, and through the simulations done under the condition that noise mismatch and the base is disturbed, it's effective and practical for RMSTCKF.

2 MULTIPLE FADING FACTOR STF

The principle of strong tracking filter(STF) is to introduce an adaptive fading factor λ_k into predicted covariance matrix $P_{k|k-1}$, thus making it real-time in adjusting gain matrix K_k online. 2 limitations of STF are:

$$\begin{cases} E[(x_k - \hat{x}_k)(x_k - \hat{x}_k)^{\mathrm{T}}] = \min & (\boldsymbol{a}) \\ E[\varepsilon_{k+i}\varepsilon_k^{\mathrm{T}}] = 0, k = 1,2,\cdots; i = 1,2,\cdots & (\boldsymbol{b}) \end{cases} \tag{1}$$

Formula (a) is the performance index that makes the filter meet the condition of minimum variance estimation. Formula (b) forces the residual sequence of filter to keep orthogonal moment by moment, conquering the problem that the performance of filer decreases when state changes suddenly or the model is unsure[4,5], thus making STF have stronger robustness and tracking capability.

2.1 Flow of RMSTCKF

Set the initial state and covariance matrix of the system are $X_{0|0}$ and P_0, and apply the third-order spherical radial cubature principle to the algorithm. The time update and measurement update based on numerical integration from article [6] are shown as following.

2.1.1 Flow of RMSTCKF

① calculate cubature points $X_{i,k-1|k-1}\left(i=1,\cdots 2n\right)$

$$X_{i,k-1|k-1}=S_{k-1|k-1}\xi_i+\hat{x}_{k-1|k-1} \tag{2}$$

Where $S_{k-1|k-1}$ is the Cholesky decomposition of $P_{k-1|k-1}$, $P_{k-1|k-1}=S_{k-1|k-1}S^{\mathrm{T}}_{k-1|k-1}$. $\xi_i=\sqrt{n}[1]_i$ denotes set of cubature points, $[1]_i\in\mathbf{R}^n$ denotes the ith element of completely symmetric point set $[1]$.

$$[1]=\left\{\begin{pmatrix}1\\0\\\vdots\\0\end{pmatrix},\begin{pmatrix}0\\1\\\vdots\\0\end{pmatrix},\cdots,\begin{pmatrix}0\\0\\\vdots\\1\end{pmatrix},\begin{pmatrix}-1\\0\\\vdots\\0\end{pmatrix},\begin{pmatrix}0\\-1\\\vdots\\0\end{pmatrix},\cdots,\begin{pmatrix}0\\0\\\vdots\\-1\end{pmatrix}\right\}$$

② calculate cubature points $X^*_{i,k/k-1}$ transmitted by state function

$$X^*_{i,k/k-1}=f\left(X_{i,k-1|k-1}\right) \tag{3}$$

③ calculate state estimation value at present moment

$$\hat{x}_{k/k-1}=\sum_{i=1}^{2n}\omega_i\left(X^*_{i,k/k-1}\right) \tag{4}$$

Where $\omega_i=1/(2n)$ is the weight value of $X^*_{i,k/k-1}$.

④ calculate predicted error covariance matrix $P^l_{k/k-1}$ without introducing fading factor

$$P^l_{k/k-1}=\sum_{i=1}^{2n}\omega_i\left(X^*_{i,k/k-1}-\hat{x}_{k/k-1}\right)\left(X^*_{i,k/k-1}-\hat{x}_{k/k-1}\right)^{\mathrm{T}}+Q_{k-1} \tag{5}$$

Mark $P^x_{k/k-1}=\sum_{i=1}^{2n}\omega_i\left(X^*_{i,k/k-1}-\hat{x}_{k/k-1}\right)\left(X^*_{i,k/k-1}-\hat{x}_{k/k-1}\right)^{\mathrm{T}}$.

2.1.2 Measurement update

⑤ calculate state cubature points $X^l_{i,k|k-1}$ after update

$$X^l_{i,k|k-1}=S^l_{k|k-1}\xi_i+\hat{x}_{k|k-1} \tag{6}$$

Where $P^l_{k/k-1}=S^l_{k|k-1}(S^l_{k|k-1})^{\mathrm{T}}$.

During initial alignment process, measurement function of system is usually linear, so it can be used to simplify the measurement update process of filter. Remark $A=\sum_{i=1}^{2n}\omega_i\left(X^l_{i,k/k-1}-\hat{x}_{k/k-1}\right)\left(X^l_{i,k/k-1}-\hat{x}_{k/k-1}\right)^{\mathrm{T}}$, and prove $A=P^l_{k/k-1}$.

From the definition of $\xi_i=\sqrt{n}[1]_i$ explained in step ①, $\xi_n\xi_n^{\mathrm{T}}$ can be expressed as:

$$\xi_n\xi_n^{\mathrm{T}}=\begin{bmatrix}0&0&\cdots&0\\0&0&&0\\\vdots&&\ddots&\vdots\\0&0&\cdots&1\end{bmatrix}_{n\times n}.$$

So $\xi_1\xi_1^{\mathrm{T}}+\xi_2\xi_2^{\mathrm{T}}+\cdots+\xi_n\xi_n^{\mathrm{T}}=E_{n\times n}$, and A can be expressed as:

$$\begin{aligned}A&=\sum_{i=1}^{2n}\omega_i\left(S^l_{k|k-1}\xi_i\right)\left(S^l_{k|k-1}\xi_i\right)^{\mathrm{T}}\\&=\omega_1S^l_{k|k-1}\xi_1\xi_1^{\mathrm{T}}(S^l_{k|k-1})^{\mathrm{T}}+\omega_2S^l_{k|k-1}\xi_2\xi_2^{\mathrm{T}}(S^l_{k|k-1})^{\mathrm{T}}+\cdots\\&\quad+\omega_{2n}S^l_{k|k-1}\xi_{2n}\xi_{2n}^{\mathrm{T}}(S^l_{k|k-1})^{\mathrm{T}}\\&=\frac{1}{n}\Big[S^l_{k|k-1}\xi_1\xi_1^{\mathrm{T}}(S^l_{k|k-1})^{\mathrm{T}}+S^l_{k|k-1}\xi_2\xi_2^{\mathrm{T}}(S^l_{k|k-1})^{\mathrm{T}}+\cdots\\&\quad+S^l_{k|k-1}\xi_n\xi_n^{\mathrm{T}}(S^l_{k|k-1})^{\mathrm{T}}\Big]\\&=\frac{1}{n}S^l_{k|k-1}(\xi_1\xi_1^{\mathrm{T}}+\xi_2\xi_2^{\mathrm{T}}+\cdots+\xi_n\xi_n^{\mathrm{T}})(S^l_{k|k-1})^{\mathrm{T}}\\&=S^l_{k|k-1}E(S^l_{k|k-1})^{\mathrm{T}}=P^l_{k/k-1}\end{aligned}$$

So $A=P^l_{k/k-1}$ is proved.

⑥ calculate cubature points $Z^l_{i,k|k-1}$ transmitted by measurement function

$$Z^l_{i,k|k-1}=H_kX^l_{i,k|k-1} \tag{7}$$

⑦ calculate measurement prediction value $\hat{z}_{k|k-1}$ at time k

$$\begin{aligned}\hat{z}_{k|k-1}&=\sum_{i=1}^{2n}\omega_iZ^l_{i,k|k-1}=H_k\sum_{i=1}^{2n}\omega_iX^l_{i,k|k-1}\\&=H_k\sum_{i=1}^{2n}\omega_i(S^l_{k|k-1}\xi_i+\hat{x}_{k|k-1})\\&=H_kS^l_{k|k-1}\sum_{i=1}^{2n}\omega_i\xi_i+H_k\sum_{i=1}^{2n}\omega_i\hat{x}_{k|k-1}\\&=\frac{H_kS^l_{k|k-1}}{2n}\left(E_{n\times n}+\left(-E_{n\times n}\right)\right)+H_k\hat{x}_{k|k-1}\\&=H_k\hat{x}_{k|k-1}\end{aligned} \tag{8}$$

It can be seen from formula (8) that the solution procedure of $\hat{z}_{k|k-1}$ makes full use of the distribution characteristics of ξ_i, without calculating $X^l_{i,k|k-1}$ in formula (6), and reducing one time cubature transformation.

⑧ calculate measurement error covariance matrix $P_{zz,k|k-1}$ and predicted cross-covariance matrix $P_{xz,k|k-1}$ both introducing fading factor at time k

Firstly, $P_{zz,k|k-1}$ and $P_{xz,k|k-1}$ without introducing fading factor can be expressed as:

$$\begin{aligned}P_{zz,k|k-1}&=\sum_{i=1}^{2n}\omega_i\left(Z_{i,k|k-1}-\hat{z}_{k|k-1}\right)\left(Z_{i,k|k-1}-\hat{z}_{k|k-1}\right)^{\mathrm{T}}+R_k\\&=\sum_{i=1}^{2n}\omega_iH_k(X^l_{i,k|k-1}-\hat{x}_{k|k-1})(X^l_{i,k|k-1}-\hat{x}_{k|k-1})^{\mathrm{T}}H_k^{\mathrm{T}}+R_k\\&=H_kAH_k^{\mathrm{T}}+R_k=H_kP^l_{k/k-1}H_k^{\mathrm{T}}+R_k\end{aligned} \tag{9}$$

$$
\begin{aligned}
P_{xz,k|k-1}^{l} &= \sum_{i=1}^{2n} \omega_i \left(X_{i,k|k-1}^{l} - \hat{x}_{k/k-1} \right)\left(Z_{i,k|k-1} - \hat{z}_{k|k-1} \right)^{\mathrm{T}} \\
&= \sum_{i=1}^{2n} \omega_i \left(X_{i,k|k-1}^{l} - \hat{x}_{k/k-1} \right)\left(H_k X_{i,k|k-1}^{l} - H_k \hat{x}_{k|k-1} \right)^{\mathrm{T}} \\
&= \left[\sum_{i=1}^{2n} \omega_i \left(X_{i,k|k-1}^{l} - \hat{x}_{k/k-1} \right)\left(X_{i,k|k-1}^{l} - \hat{x}_{k|k-1} \right)^{\mathrm{T}} \right] H_k^{\mathrm{T}} \\
&= A H_k^{\mathrm{T}} = P_{k/k-1}^{l} H_k^{\mathrm{T}}
\end{aligned} \tag{10}
$$

And use 2 multiple fading factor matrix introduced in $P_{k/k-1}^{l}$ to replace single fading factor, the form is:

$$P_{k/k-1} = \lambda_k P_{k/k-1}^{x} \lambda_k' + Q_{k-1} \tag{11}$$

So $P_{zz,k|k-1}$ and $P_{xz,k|k-1}$ that introduce fading factor can be expressed as:

$$P_{zz,k|k-1} = H_k P_{k/k-1} H_k^{\mathrm{T}} + R_k \tag{12}$$

$$P_{xz,k|k-1} = P_{k/k-1} H_k^{\mathrm{T}} \tag{13}$$

⑨ calculate filter gain matrix K_k

$$\hat{x}_{k|k} = \hat{x}_{k|k-1} + K_k \left(z_k - \hat{z}_{k|k-1} \right) \tag{14}$$

⑩ filter state value at current moment

$$\hat{x}_{k|k} = \hat{x}_{k|k-1} + K_k \left(z_k - \hat{z}_{k|k-1} \right) \tag{15}$$

$$P_{k|k} = P_{k|k-1} - K_k P_{zz,k|k-1} (K_k)^{\mathrm{T}} \tag{16}$$

From the algorithm flow above, it can be seen that the proposed RMSTCKF in this article can get the solution of $P_{k/k-1}$, $P_{zz,k|k-1}$ and $P_{xz,k|k-1}$ only by finishing one time cubature transformation, thus making the calculating quantity reduced and making the result satisfy the limitations in formula (1), completing the whole STF afterwards.

2.2 Sub-optimal solution of multiple fading factor

The sub-optimal solution of multiple fading factor is[2]:

$$\lambda_k = diag[\lambda_k^1 \ \lambda_k^2 \ \cdots \ \lambda_k^n] \tag{17}$$

$$\lambda_k^i = \max\left(1, \frac{\alpha_i \mathrm{tr}[N_k]}{\sum_{i=1}^{n} \alpha_i M_k^{ii}} \right) \tag{18}$$

$$N_k = V_k - H_k Q_{k-1} H_k^{\mathrm{T}} - R_k \tag{19}$$

$$M_k = F_{k|k-1} P_{k-1|k-1} F_{k|k-1}^{\mathrm{T}} H_k^{\mathrm{T}} H_k \tag{20}$$

$$e_k = z_k - H_k \hat{x}_{k/k-1} \tag{21}$$

$$V_k = \begin{cases} e_1 e_1^{\mathrm{T}} & k = 0 \\ \dfrac{\rho V_{k-1} + e_k e_k^{\mathrm{T}}}{1+\rho} & k > 0 \end{cases} \tag{22}$$

Where $P_{k/k-1}^{x}$ denotes the state predicted error covariance matrix without considering λ_k and v_{k-1}, $P_{k/k-1}^{l}$ denotes the state predicted error covariance matrix considering λ_k and v_{k-1}, $P_{k/k-1}^{l} = P_{k/k-1}^{x} + Q_{k-1}$; $P_{k/k-1}^{xz,l}$ denotes cross-covariance matrix without considering λ_k ; $X_{i,k|k-1}^{l}$ denotes cubature points produced by one-step prediction $\hat{x}_{k/k-1}^{l}$ without considering λ_k; N_k, M_k, H_k and $V_{0,k}$ are parameter matrix; weakening factor $\beta = 3$, forgetting factor $\rho = 0.95$.

From article [2] we know that $P_{k/k-1}^{x}$ and H_k can be equally expressed as:

$$
\begin{aligned}
P_{k/k-1}^{x} &= F_{k|k-1} P_{k-1|k-1} F_{k|k-1}^{\mathrm{T}} \\
&= \sum_{i=1}^{m} \omega_i \left(X_{i,k/k-1}^{*} - \hat{x}_{k/k-1} \right)\left(X_{i,k/k-1}^{*} - \hat{x}_{k/k-1} \right)^{\mathrm{T}}
\end{aligned} \tag{23}
$$

Because H_k is linear measurement matrix, and M_k, N_k can be expressed as:

$$M_k = P_{k/k-1}^{x} H_k^{\mathrm{T}} H_k \tag{24}$$

$$N_k = V_k - H_k Q_{k-1} H_k^{\mathrm{T}} - R_k \tag{25}$$

Put formula (23), (24), (25) into formula (19), (20), the sub-optimal solution of multiple fading factor can be get.

3 Large azimuth misalignment angle error model

Set inertial coordinate system is i-frame, earth coordinate system is e-frame, carrier coordinate system is b-frame. Select local geography coordinate system to be navigation coordinate system n-frame, the navigation coordinate system calculated by SINS is p-frame. Due to the influence of different error origin of SINS, the p and n calculated by navigation computer have misalignment angle error $\phi = [\phi_x \ \ \phi_y \ \ \phi_z]^{\mathrm{T}}$, which denotes a group of euler angle that rotate from n-frame to p-frame, the rotation order is defined as ϕ_z, ϕ_x, ϕ_y. So the coordinate transfer matrix C_n^p from n-frame to p-frame is:

$$
\begin{aligned}
C_n^p &= \begin{bmatrix} \cos\phi_y & 0 & -\sin\phi_y \\ 0 & 1 & 0 \\ \sin\phi_y & 0 & \cos\phi_y \end{bmatrix} \cdot \begin{bmatrix} 1 & 0 & 0 \\ 0 & \cos\phi_x & \sin\phi_x \\ 0 & -\sin\phi_x & \cos\phi_x \end{bmatrix} \\
&\quad \cdot \begin{bmatrix} \cos\phi_z & \sin\phi_z & 0 \\ -\sin\phi_z & \cos\phi_z & 0 \\ 0 & 0 & 1 \end{bmatrix} \\
&= \begin{bmatrix} \cos\phi_y \cos\phi_z + \sin\phi_y \sin\phi_x \sin\phi_z & -\cos\phi_y \sin\phi_z + \sin\phi_y \sin\phi_x \cos\phi_z & -\sin\phi_y \cos\phi_x \\ \cos\phi_x \sin\phi_z & \cos\phi_x \cos\phi_z & \sin\phi_x \\ \sin\phi_y \cos\phi_z - \cos\phi_y \sin\phi_x \sin\phi_z & -\sin\phi_y \sin\phi_z - \cos\phi_y \sin\phi_x \cos\phi_z & \cos\phi_y \cos\phi_x \end{bmatrix}
\end{aligned} \tag{26}
$$

3.1. SINS attitude error equation

ω_{np}^{p} is the projection of p to the angular velocity on n, the differential equation of EPEA is[1]:

$$\dot{\phi} = C_{\phi}^{-1} \cdot \omega_{np}^{p} \tag{27}$$

$$C_{\phi}^{-1} = \frac{1}{\cos\phi_x}\begin{bmatrix} \cos\phi_x\cos\phi_y & 0 & \sin\phi_y\cos\phi_x \\ \sin\phi_x\sin\phi_y & \cos\phi_x & -\sin\phi_x\cos\phi_y \\ -\sin\phi_y & 0 & \cos\phi_y \end{bmatrix} \tag{28}$$

From the derivation of ω_{np}^{p}, it can be expressed as $\omega_{np}^{p} = \left(I - C_n^p\right)\omega_{in}^{n} + \tilde{\omega}_{in}^{n} - C_n^p\tilde{\omega}_{ib}^{b}$, then put ω_{np}^{p} in formula (27), ignore the second-order and higher-order value, and SINS attitude error function is expressed as:

$$\dot{\phi} = C_{\phi}^{-1}\left(\left(I - C_n^p\right)\tilde{\omega}_{in}^{n} + C_n^p\delta\omega_{in}^{n} - C_b^p\delta\omega_{ib}^{b}\right) \tag{29}$$

Where C_b^p is the coordinate transfer matrix from the carrier (b-frame) to computed navigation coordinate(p-frame), C_n^p is the coordinate transfer matrix from the ideal navigation coordinate (n-frame) to computed navigation coordinate (p-frame), $\tilde{\omega}_{in}^{n}$ denotes the angular rate of the b-frame with respect to n-frame expressed in b-frame, $\delta\omega_{in}^{n}$ is calculation error of navigation coordinate, $\delta\omega_{ib}^{b}$ is measurement error of gyro.

3.2 SINS nonlinear filter model

Assuming that $\delta\omega_{ib}^{b}$ is composed of gyro constant drift $\varepsilon_b = \left[\varepsilon_{bx}\ \varepsilon_{by}\ \varepsilon_{bz}\right]^{\mathrm{T}}$ and gyro random drift $\omega_g = \left[\omega_{gx}\ \omega_{gy}\ \omega_{gz}\right]^{\mathrm{T}}$, ω_g is Gaussian white noise with zero mean. δf^b is composed of accelerometer constant biases $\nabla_b = \left[\nabla_{bx}\ \nabla_{by}\ \nabla_{bz}\right]^{\mathrm{T}}$ and accelerometer random drift $\omega_a = \left[\omega_{ax}\ \omega_{ay}\ \omega_{az}\right]^{\mathrm{T}}$, ω_a is Gaussian white noise with zero mean.

Set the state vector of SINS $X = [\phi_x\ \phi_y\ \phi_z\ \delta v_x\ \delta v_y\ \nabla_{bx}\ \nabla_{by}\ \varepsilon_{bx}\ \varepsilon_{by}\ \varepsilon_{bz}]^{\mathrm{T}}$, specific expression of state matrix $f(x)$ can refer to formula (27), (28), (29), (32), state noise vector is $w = [w_{gx}\ w_{gy}\ w_{gz}\ w_{ax}\ w_{ay}\ 0\ 0\ 0\ 0\ 0]^{\mathrm{T}}$. The observation value is velocity error, the observation vector is $Z = \left[\delta v_x\ \delta v_y\right]^{\mathrm{T}}$, observation matrix is $H = \left[I_{2\times2}\ \ 0_{2\times8}\right]$, observation noise vector is v that is Gauss white noise with zero mean. So the nonlinear filter model of SINS can be expressed as:

$$\begin{cases} \dot{X} = f(X) + w \\ Z = HX + v \end{cases} \tag{33}$$

When the initial alignment is operated with stationary base, the information of longitude, latitude and position is precisely known, and there is no relative motion between n-frame and e-frame, so $v^n = 0$, $\omega_{en}^{n} = 0$, $\omega_{ie}^{n} = [0\ \ \omega_{ie}\cos L\ \ \omega_{ie}\sin L]^{\mathrm{T}}$, where L denotes geography latitude, $f^n = C_b^n f^b = [0\ 0\ -g]^{\mathrm{T}}$, so the nonlinear initial alignment model of stationary base can be get on the basis of formula (29) and (32).

$$\begin{cases} \dot{\phi} = C_{\phi}^{-1}\left(I - C_n^p\right)\omega_{ie}^{n} - C_{\phi}^{-1}C_b^p\varepsilon^b + C_{\phi}^{-1}C_b^p w_g^b \\ \delta\dot{v}^n = (C_n^p - I)C_b^n f^b + C_b^p\nabla^b + C_b^p w_a^b \\ \dot{\varepsilon}^b = 0 \\ \dot{\nabla}^b = 0 \end{cases} \tag{34}$$

Where the coefficient matrix C_{ϕ}^{-1} can be approximated as:

$$C_{\phi}^{-1} = \begin{bmatrix} 1 & 0 & \phi_y \\ 0 & 1 & -\phi_x \\ -\phi_y & 0 & 1 \end{bmatrix} \tag{35}$$

Where $\sin\phi_x \approx \phi_x$, $\cos\phi_x \approx 1$, $\sin\phi_y \approx \phi_y$, $\cos\phi_x \approx 1$.

4 Simulation

Apply RCKF, RSTCKF and RMSTCKF to large azimuth misalignment angle model to carry on I nitial alignment. The position of SINS is 45 ° N, 108 ° E, height equals to 100m, the initial state of system is $X_0 = 0$, and the parameters of inertial measurement unit are shown in table 1.

Table 1. Main parameters of inertial measurement unit

Parameters of gyro and accelerator	Value of parameter
Gyro constant drift /((°) / h)	0.1
Gyro random drift /((°) / h)	0.05
Dynamic range of gyro /((°) / s)	±200
Accelerator constant drift /μg	100
Accelerator random drift /μg	50
Dynamic range of accelerator / g	±30

Set initial value as:

$$P(0) = \mathrm{diag}[(2°)^2\ (2°)^2\ (10°)^2\ (0.1\mathrm{m/s})^2\ (0.1\mathrm{m/s})^2\ (100ug)^2\ (100ug)^2 (0.1°)^2\ (0.1°)^2\ (0.1°)^2\ ]$$

$$Q = diag[(0.05°/h)^2\ (0.05°/h)^2\ (0.05°/h)^2\ (50\mu g)^2\ (50\mu g)^2\ (50\mu g)^2\ 0_{5\times1}]$$

$$R = diag[(0.1\mathrm{m/s})^2\ (0.1\mathrm{m/s})^2]$$

4.1 Initial alignment simulation on stationary base

Set the initial misalignment angle of E-direction, N-direction and S-direction is $\phi = [1°\ 1°\ 30°]^{\mathrm{T}}$, apply above 3 algorithms to initial alignment simulation for 600s, the azimuth misalignment angle error is shown in figure 1. From the aspect of convergence, there's little difference between RCKF, RSTCKF and RMSTCKF. Through 100 times Monte-Carlo simulation, the accuracy of initial alignment is measured by the RMSE of azimuth misalignment angle at last 100s, and the specific results are shown in table 2. From the aspect of convergence speed, RMSTCKF converges around limited accuracy at 120s, faster than RSTCKF and RCKF. System isn't affected by model error and external disturbance in ideal environment, residual sequence keeps orthogonal moment by moment, and the above 3 algorithms can keep better filter capability.

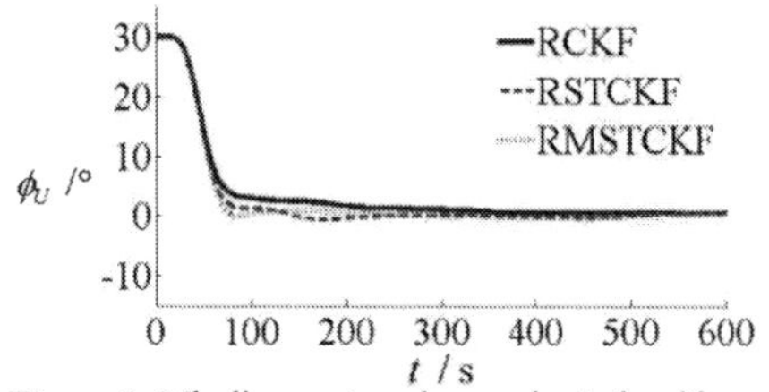

Figure 1. Misalignment angle error by 3 algorithms

Table 2. Alignment results by 3 algrtithms

Misalignment angle error	RCKF	RSTCKF	RMSTCKF	Limited accuracy(′)
Phi-E	-0.383	-0.390	0.386	-0.3438
Phi-N	0.413	0.418	0.425	0.3440
Phi-U	37.690	37.343	37.181	-32.4000

4.2 Initial alignment simulation under the condition that system noise variance matrix mismatch

In order to prove the adaptability and robustness of RMSTCKF proposed in this article, expand true noise covariance matrix of system by nine times, the set value of Q in filter model is invariant, and other simulation conditions are the same. The misalignment angle curve in S-direction is shown in figure 2. When system noise mismatches, 3 algorithms all converge, and the filter value at 600s are respectively 192.36', 39.12' and 38.64'. The wave of RCKF is the most obvious, and the wave gradually disappears after 400s, the convergence speed and accuracy of RCKF is bad. Performance of RSTCKF and RMSTCKF are better than RCKF obviously, but there's still small disturbance in RSTCKF, and both of them can converge to limited accuracy quickly.

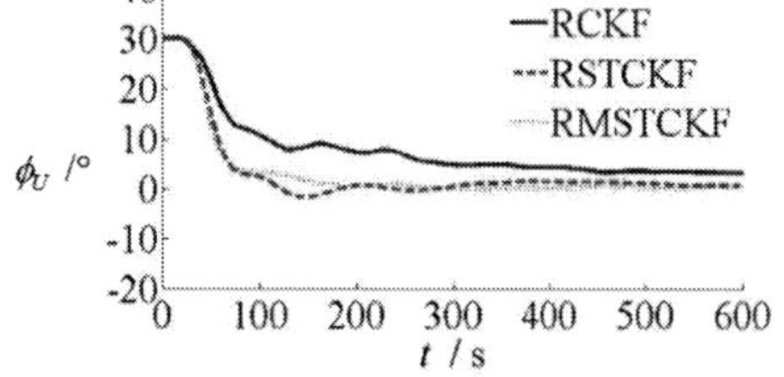

Figure 2. Misalignment angle error of 3 algorithms under the condition that system noise mismatch

4.3 Initial alignment simulation under the condition that base is disturbed

Stationary base initial alignment simulation of SINS doesn't consider practical factors of practical working like external disturbance. For the purpose of checking the robustness and tracking ability of RMSTCKF while facing external disturbance, when alignment time reaches 40s, add sine swaying disturbance which lasts 10s to the origin of base.

$$\begin{cases} \theta = 5°\sin\left(\frac{2\pi}{10}(t-40)\right) \\ \gamma = 10°\sin\left(\frac{2\pi}{10}(t-40)\right) \end{cases} \quad 40 < t < 50 \tag{36}$$

Other simulation conditions are invariant, carry on the initial alignment simulation that last 600s with above 3 algorithms. Misalignment angle error curve at S-direction is shown in figure 3, figure 3 (a) and figure 3 (b) are enlarged detail where is disturbed.

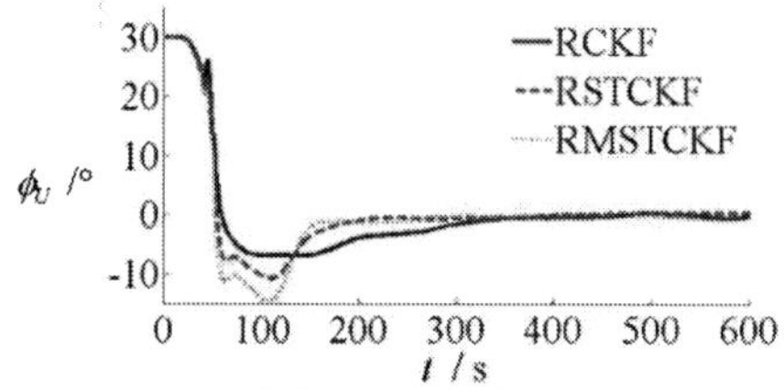

Figure (a). Whole curve of misalignment angle error under 3 algorithms

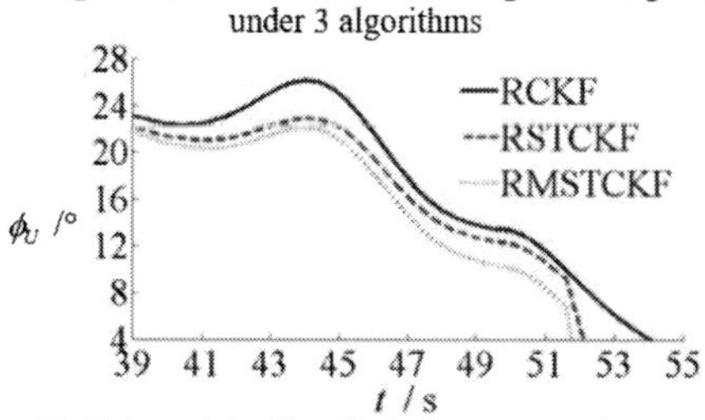

Figure (b). Enlarged detail s of misalignment angle error under 3 algorithms

Figure 3. Misalignment angle error of 3 algorithms while there's sudden disturbance

It can be seen from figure 3 (a) that RMATCKF and RSTCKF have stronger adjustable ability and shorter convergence time than RCKF when system is disturbed. RMSTCKF converges to limited accuracy around 150s, RSTCKF converges to limited accuracy around 220s, and RCKF converges to limited accuracy around 500s.

It can be seen from figure 3 (b) that the filter value of 3 algorithm have different degree of wave due to the disturbance during 40s~50s. However, the disturbance wave of RMSTCKF is obviously smaller than the other 2 algorithms. The disturbance wave of RSTCKF is bigger than RMSTCKF but smaller than RCKF, and the disturbance wave of RCKF is the biggest that filter performance of RCKF is obviously decreasing, being disable to meet the alignment requirements of SINS. From above analysis, it's obvious that the proposed RMSTCKF has better adaptability and higher filter accuracy, equipping with stronger robustness and tracking capability when state changes suddenly.

5 Conclusion

An improved algorithm RMSTCKF which is applicable for initial alignment of large azimuth misalignment angle is propose in this paper, the calculating flow and sub-optimal solution of multiple fading factor matrix are provided. The nonlinear error equation of SINS expressed by EPEA is derived. Complete the initial alignment simulation of large azimuth misalignment angle with 3 algorithms (RCKF, RSTCKF and RMSTCKF) under 3 different conditions that include stationary base, system noise covariance mismatch and the base is disturbed. Results show that RMSTCKF proposed in this paper has stronger adaptability, performing stronger robustness and tracking ability. Compared with the other 2 algorithms, RMSTCKF is better in practical engineering.

References

1. YAN Gong-min, YAN Wei-Sheng, XU De-min. Application of simplified UKF in SINS initial alignment for large misalignment angles [J]. Journal of Chinese Inertial Technology, 3, **16** (2008)
2. DU Zhan-long, Li Xiao-min. Fault estimation and prediction based on multiple fading factors strong tracking UKF and constrained AR model [J]. Control and Decision, 9, **29** (2014)
3. XUE Hai-jian, Wang Jie, Guo Xiao-song, et al. Design of a Strong Tracking UKF for Nonlinear Self-Alignment of SINS [J]. Journal of Shang Hai Jiao Tong University, 9, **49** (2015)
4. Zhou D H, FRANK P M. Strong tracking filtering of nonlinear time-varying stochastic systems with colored noise Application to parameter estimation and empirical robustness analysis [J]. International Journal of Control, 2, **65** (1996)
5. Wang L, Wu L, Guan Y, et al. Online sensor fault detection based on an improved strong tracking filter [J]. Sensors, 2, **15** (2015)
6. ZHANG Wen-Jie, Wang Shi-yuan, Feng Ya-li, et al, Huber-based high-degree cubature Kalman tracking algorithm [J]. Acta Phys. Sin., 8, **65** (2016)
7. HU Gao-Ge, Liu Yi-han, Gao Du-Sheng, et al. Improved strong tracking UKF and its application in INS/GPS integr-ated navigation [J]. Journal of Chinese Inertial Technology, 5, **22** (2014)
8. Jia B, Xin M, Cheng Y. High-degree cubature Kalman filter [J]. Automatica, **49** (2013)
9. Pei H L, Sanjeev A, Tharaka A, et al. A Gaussian-sum based cubature Kalman filter for bearings only tracking [J]. IEEE Transactions on Aerospace and Electronic Systems, 2, **49** (2013)
10. Lin X G, Xu S S, Xie Y H, Multi-sensor hybrid fusion algorithm based on adaptive square root cubature Kalman filter [J]. Journal of Marine Science and Application, **12** (2013)

Economic Dispatch Solution Using Moth-Flame Optimization Algorithm

Mohd Herwan Sulaiman[1], Zuriani Mustaffa[2], Muhammad Ikram Mohd Rashid[1] and Hamdan Daniyal[1]

[1]*Faculty of Electrical & Electronics Engineering, Universiti Malaysia Pahang, 26600 Pekan Pahang, Malaysia*
[2]*Faculty of Computer Systems and Software Engineering, Universiti Malaysia Pahang, 26300Gambang Pahang, Malaysia*

Abstract. This paper proposes an application of a recent nature inspired optimization technique namely Moth-Flame Optimization (MFO) algorithm in solving the Economic Dispatch (ED) problem. In this paper, the practical constraints will be included in determining the minimum cost of power generation such as ramp rate limits, prohibited operating zones and generators operating limits. To show the effectiveness of proposed algorithm, two case systems are used: 6-units and 15-units systems and then the performance of MFO is compared with other techniques from literature. The results show that MFO is able to obtain less total cost than those other algorithms.

1 Introduction

Economic Dispatch (ED) is one of the complex problems in power system operation that draws a lot of attention of many researchers. The task of ED is to find the optimal power generation to meet the demand at the minimum cost by fulfilling all the constraints. In addition, by taking the practical and operational constraints into account, it makes ED highly nonlinear constrained optimization problem, especially for larger systems. Any improvements in optimal output scheduling can contribute significantly in terms of cost savings for a long run.

To date, there are massive techniques and algorithms that have been applied to solve the ED problems. The application of nature inspired algorithms becoming the popular choices of many researchers that have been proven through literatures. The implementation of Particle Swarm Optimization (PSO) into ED solution has been proposed in [1]. The variants of Genetic Algorithm (GA) namely Beeder GA (BGA), Fast Navigating GA (FNGA), Twin Removal GA (TRGA), Kite GA (KGA) and United GA 9UGA) have been proposed in [2] to solve ED problems. Nevertheless, the results presented in [2] still unsatisfactory if compared with the Backtracking Search Algorithm that has been proposed in [3].

Ref. [4] has proposed Cuckoo Search Algorithm (CSA) and the Firefly Algorithm (FA) has been proposed in [5] to solve ED problems. The modification of FA (MFA) based on [5] has been proposed in [6] to solve ED problems with practical constraints. Other algorithms that also have been proposed in literature to solve ED are such as Grey Wolf Optimizer (GWO) [7], Flower Pollination Algorithm (FPA) [8], Whale Optimization Algorithm (WOA) [9], Improve/ modified PSO [10, 11] and Hybrid Bee Algorithm with Tabu Search (BA/TS) [12].

In this paper, the recent developed algorithm namely Moth-Flame Optimization (MFO) algorithm has been applied to solve practical ED problems. MFO is inspired by moth's navigation at night proposed by [13]. The rest of the paper is organized as follows. Section 2 discuss the ED problem formulation followed by the MFO algorithm in Section 3. Section 4 presents the implementation of MFO in solving ED problem and Section 5 presents the results and discussion. Finally, the conclusion is stated in Section 6.

2 Economic Dispatch

Economic Dispatch (ED) is about to find the optimal power generation at the lowest cost to reliably serve the loads subject to the constraints. In ED, the cost function can be mathematical described as the quadratic function, as follows:

$$F_i(P_i) = \alpha_i P_i^2 + \beta_i P_i + \gamma_i \quad (1)$$

where α_i, β_i and γ_i are the coefficients of generator i. The objective function to be minimized are the summation of cost function in (1), which can be expressed as follows:

$$\min f = \sum_{i=1}^{n} F_i(P_i) \quad (2)$$

where n is the number of generator units. Eqns. (1) and (2) are subject to the power balanced equation as equality constraint to be satisfied where can be expressed as follows:

$$\sum_{i=1}^{n}(P_i) = P_{Demand} + P_{Loss} \quad (3)$$

where P_{Demand} is the total load demand, P_{Loss} is the total loss of the power system. Since the losses in the interconnected power system cannot be avoided, the *B-coefficient* method is used to calculate the losses apart of load flow calculation, as follows:

$$P_{Loss} = \sum_{i=1}^{n}\sum_{j=1}^{n} P_i B_{ij} P_j + \sum_{i=1}^{n} B_{0i} P_i + B_{00} \quad (4)$$

For the inequality constraint, the power generation must be operated within its individual limits, as follows:

$$P_{i(\min)} \leq P_i \leq P_{i(\max)} \qquad i = 1, 2, \ldots, n \quad (5)$$

Other than that, by considering the practical constraints, the operating range of all generation units are restricted by their ramp rate limits [14], which are expressed as follow:

$$P_i - P_i^o \leq UR_i \quad \text{if generation is increased} \quad (6)$$

$$P_i^o - P_i \leq DR_i \quad \text{if generation is decreased} \quad (7)$$

where P_i^o is the previous power generation of unit *i*. UR_i and DR_i are the up-ramp and down-ramp limits in MW/h of the *i*-th generator, respectively. The generator operation constraints with the ramp rate limit now becomes as follows:

$$\max\left(P_i^{\min}, UR_i - P_i^o\right) \leq P_i \leq \min\left(P_i^{\max}, P_i^o - DR_i\right) \quad (8)$$

The other practical constraint that will be considered in this paper is the prohibited operating zones for the generators. The condition of this situation is visually shown in Fig. 1. Since the shape of the input-output curve in the neighborhood of the prohibited zones are difficult to determined, the best economical way is by avoiding the operation in these areas. The feasible operating zones of a unit can be determined as follow [15]:

$$\begin{cases} P_i^{\min} \leq P_i \leq P_{i,1}^{lower} \\ P_{i,j-1}^{upper} \leq P_i \leq P_{i,j}^{lower}, j = 2, 3, \ldots, PZ_i \\ P_{i,PZ_i}^{upper} \leq P_i \leq P_i^{\max} \end{cases} \quad (9)$$

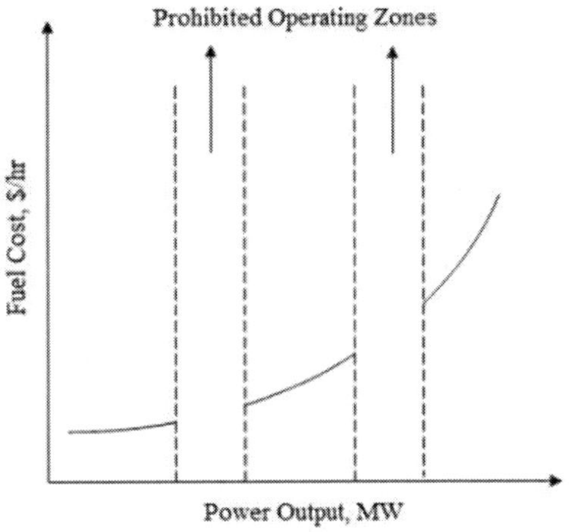

Figure 1. Fuel cost curve with prohibited zone

3 Moth-Flame Optimization Algorithm

MFO algorithm is inspired by the moth's special navigation at night proposed by [13]. Moths fly at night by referring to the moon light as orientation for navigation. They fly by maintaining a fixed angle with respect to the moon for straight path. Moths are tricked by the artificial light which make them fly spirally around the lights. However, the effort of moths to maintain a similar angle to the light source which is extremely close compared to the moon will causes a deadly spiral fly path for them. This behavior is depicted in the Fig. 2.

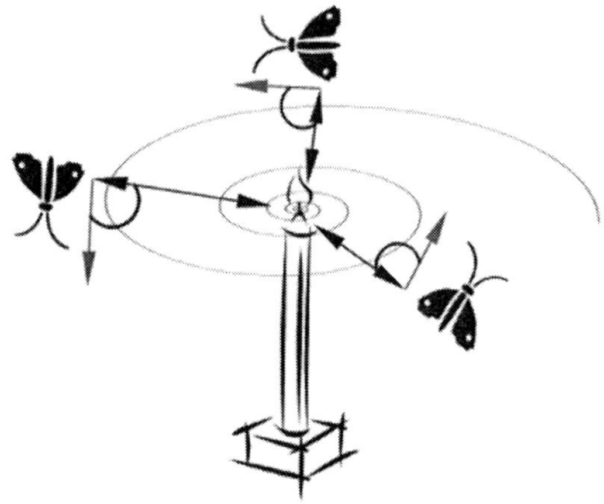

Figure 2. Spiral flying path around close light sources [13].

The development of MFO initially can be described as the following expression:

$$M = \begin{bmatrix} m_{1,1} & m_{1,2} & \cdots & m_{1,d} \\ \vdots & \vdots & \cdots & \vdots \\ \vdots & \vdots & \cdots & \vdots \\ m_{n,1} & m_{n,2} & \cdots & m_{n,d} \end{bmatrix} \quad (10)$$

where M is a set of moths, n is the number of moths and d is the number of variables (dimension). It is assumed that the candidate of solutions is moths. Other key components in MFO of course are the flames which are constituted also in the form of matrix as follows:

$$F = \begin{bmatrix} F_{1,1} & F_{1,2} & \cdots & F_{1,d} \\ \vdots & \vdots & \cdots & \vdots \\ \vdots & \vdots & \cdots & \vdots \\ F_{n,1} & F_{n,2} & \cdots & F_{n,d} \end{bmatrix} \tag{11}$$

where n and d are the number of moth and dimension respectively. Both (10) and (11) are assumed to store the corresponding fitness values in an array as expressed below:

$$OM = \begin{bmatrix} OM_1 \\ \vdots \\ \vdots \\ OM_n \end{bmatrix} \tag{12}$$

$$OF = \begin{bmatrix} OF_1 \\ \vdots \\ \vdots \\ OF_n \end{bmatrix} \tag{13}$$

From (12) and (13), both moth and flame are solutions. The difference between them is how to treat and update them for each iteration. In MFO, the moths are actual search agents that move around the search space, whereas flames are the best position of moths so far. Thus, flames can be treated as flags that are dropped by moths when searching the search space. In this mechanism, a moth never loses its best solution [13].

The position of each moth is updated with respect to the flame by using the following expression:

$$M_i = S(M_i, F_j) \tag{14}$$

where M_i indicates the i-th moth, F_j is the j-th flame and S is the spiral function. A logarithmic spiral function is selected as the main update mechanism of moth, such as [13]:

$$S(M_i, F_j) = D_i \cdot e^{bt} \cdot \cos(2\pi t) + F_j \tag{15}$$

where D_i is the distance of the i-th moth for the j-th flame, b is a constant for defining the shape of the logarithmic spiral and t is a random number between -1 and 1. To prevent from trapped in local optima, each moth is obliged to update its position using one flame only in eqn. (15). At each iteration and after updating the flames, the flames are sorted based on the fitness values. Then the moths are updating their position with respect to the corresponding flames. Details description of MFO can be obtained in [13].

4 MFO for ED Problem

The implementation of MFO in solving the ED problem is by obtaining the optimal values power generation while satisfying all the constrained mentioned in section 2. Initially, the number of moths or search agents and maximum iteration are set. The population (candidate for solution) is constructed in matrix form as depicted in eqn. (10) where the row represents the number of moths and the column represents the number of control variables to be optimized (power generation).

To obtain the objective function, each position of moth is evaluated to obtain the total cost using (2). Once the total cost has obtained for respected moth (after updating with the flames such in eqn. (11-14)), the matrix is sorted where the best solution so far is located at the top while the worst result is located at the bottom of the population matrix. If the updated variables are out of bound from the constraints, they are pegged at the minimum or maximum boundaries so that the result obtained is correct. To deal with the equality constraint such in (3), the penalty method was used. The penalty is reflected to the power balance mismatch and embedded in the cost function (2) as follows:

$$F = (F) + PF * abs\left[\left(\sum_{i=1}^{n} P_i\right) - P_{Demand} - P_{Loss}\right] \tag{16}$$

where PF is the penalty factor. The implementation of MFO in solving ORPD is exhibited in Fig. 3.

5 Results and Discussion

To show the effectiveness of proposed MFO in solving ED problem, 2 test systems are used: 6-units and 15-units generation systems. The simulation was implemented in MATLAB.

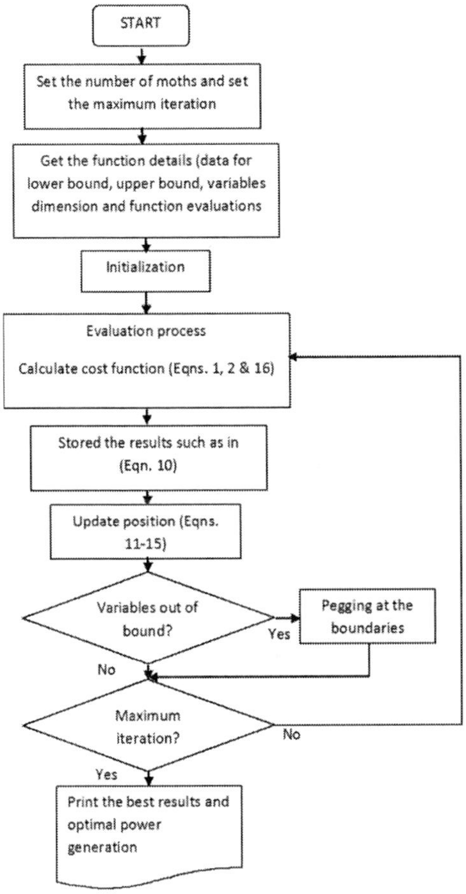

Figure 3. Flow of proposed MFO for solving ED problem

5.1 6-units system

The system consists of 6 thermal units in 26 bus system including 46 transmission lines. The load demand is set to 1263 MW. The characteristics of this system can be obtained in [6] and the B-loss coefficients can be obtained in [1].

The best results of MFO with other methods such as Kite Genetic Algorithm (KGA) [2], PSO [1], Backtracking Search Algorithm (BSA) [3] and Modified Firefly Algorithm (MFA) [6] are tabulated in Table I. From this table, it can be seen that the proposed MFO has a better solution in terms of total cost generated compared to others. MFO also gave the best results in terms of minimum line loss which is 12.414 MW.

Table 1. Best ED Solution for 6-units system

Generator Output	KGA	PSO	BSA	MFA	MFO
P1	447.66	447.5	447.4902	445.08	446.71
P2	173.60	173.32	173.3308	173.08	173.22
P3	262.91	263.47	263.4559	264.42	263.96
P4	139.18	139.06	139.0602	139.59	139.2
P5	165.58	165.48	165.4804	166.02	165.68
P6	87.03	87.13	87.1409	87.21	86.64
Line Loss	12.96	12.958	12.9583	12.41	12.414
Cost ($/h)	15449.89	15,450	15,449.89	15,443	15442.66

To analyze the performance of MFO in optimizing the power generation for this system, various number of search agents are used: 20, 30 and 40. Fig. 4 shows the performance in terms of total generation cost ($/hr) versus free runs for various numbers of search agents of MFO. It can be noted that the results of 20 search agents are slightly better compared to others in obtaining the minimum cost which is at the run # 24 which is the best cost of $ 15,442.66/ hr. However, 30 and 40 search agents are quite stable and consistent compared to 20 search agents' solutions. Fig. 5 shows the convergence curve of the best results of each number of search agents among 30 free running simulations. It can be noted that all search agents are converged less than 70 iterations.

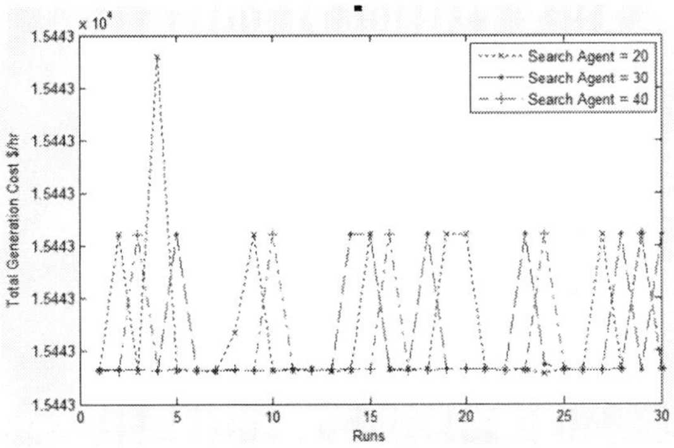

Figure 4. Performance for 20, 30 and 40 search agents for 30 free running simulations for 6-units system.

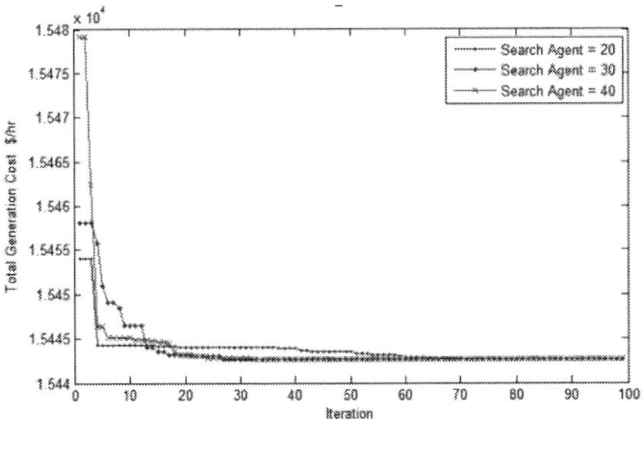

Figure 5. Convergence curve of 20, 30 and 40 search agents for 6-units system.

5.2 15-units system

For this system, the load demand is set to 2630. The generators' characteristics can be obtained in [1]. The prohibited operating zones embedded in the units 2, 5, 6 and 12. A detailed results of the optimal ED solution for this system of MFO together with Twin Removal Genetic Algorithm (TRGA) [2], BSA [3], FA [5] and Cuckoo Search Algorithm (CSA) [4] are tabulated in Table II. It can be noted that MFO gave the best results in terms of minimum cost of generation: $ 32, 687.02/hr which is less $0.02/hr compared to CSA [4]. However, the total loss obtained by MFO is better where MFO produces 30.0522 MW compared to CSA, which is 30.0541 MW.

Again, in order to analyze the performance of MFO in optimizing the power generation for this system, 20, 30 and 40 search agents are used. Fig. 6 shows the convergence curve of the best results of each number of search agents among 30 free running simulations. It can be concluded that all search agents are converged less than 100 iterations. Fig. 7 shows the performance in terms of total generation cost ($/hr) versus free runs for various numbers of search agents of MFO. It can be seen that the results of 30 search agents are slightly better compared to others in obtaining the minimum cost which is at the run # 17 which is the best cost of $ 32,687.02/ hr. Nevertheless, 40 search agents gave the consistent results throughout 30 free running simulations.

TABLE 2. BEST ED SOLUTION FOR 15-UNITS SYSTEM

Unit	TRGA (MW)	BSA (MW)	FA MW)	CSA (MW)	MFO (MW)
1	455.00	455.00	454.97	455.00	455.00
2	380.00	380.00	380.00	380.00	380.00
3	130.00	130.00	130.00	130.00	130.00
4	130.00	130.00	130.00	130.00	130.00
5	169.97	170.00	169.80	170.00	170.00
6	460.00	460.00	460.00	460.00	460.00
7	430.00	430.00	430.00	430.00	430.00
8	69.38	71.64	86.89	84.94	80.78
9	61.30	59.02	47.26	44.22	48.36
10	159.97	160.00	155.20	160.00	160.00
11	80.00	80.00	79.87	80.00	80.00
12	80.00	80.00	79.93	80.00	80.00
13	25.03	25.00	25.00	25.00	25.00
14	15.00	15.00	15.05	15.00	15.00
15	15.00	15.00	15.22	15.00	15.00
Loss (MW)	30.65153	30.6609	30.1329	30.0541	**30.0522**
Cost ($/h)	32,704.53	32,704.45	32,689.39	32687.04	**32,687.02**

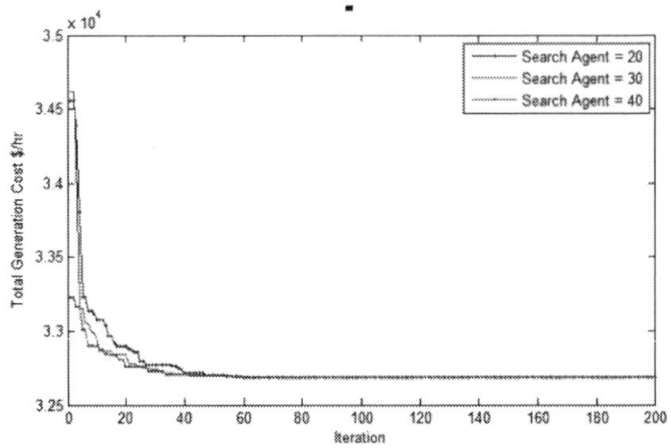

Figure 6. Convergence curve of 20, 30 and 40 search agents for 6-units system.

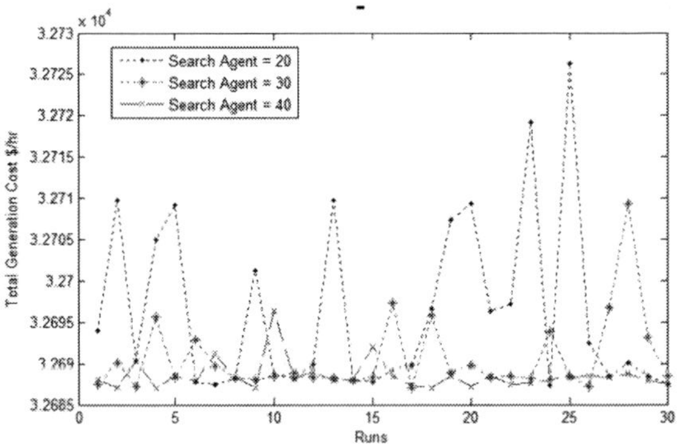

Figure 7. Performance for 20, 30 and 40 search agents for 30 free running simulations for 15-units system.

6 Conclusion

This paper has proposed a recent nature inspired computing algorithm, Moth-Flame Optimization algorithm in solving economic dispatch problem. The effectiveness of MFO was demonstrated using 6-units and 15-units systems. Simulation results showed that MFO is better compared to other selected algorithms in terms of finding the minimum cost of generation in $/hr. The implementation of MFO into economic emission dispatch will be proposed in the near future.

Acknowledgment

This research study is supported by Ministry of Higher Education Malaysia (MoHE) and Universiti Malaysia Pahang under Fundamental Research Grant Scheme #RDU 170129.

References

1. G. Zwe-Lee, "Particle swarm optimization to solving the economic dispatch considering the generator constraints," *IEEE Transactions on Power Systems,* vol. 18, pp. 1187-1195, 2003.
2. S. Rastgoufard, S. Iqbal, M. T. Hoque, and D. Charalampidis, "Genetic algorithm variant based effective solutions for economic dispatch problems," in *2018 IEEE Texas Power and Energy Conference (TPEC)*, 2018, pp. 1-6.

3. M. Modiri-Delshad, S. H. Aghay Kaboli, E. Taslimi-Renani, and N. A. Rahim, "Backtracking search algorithm for solving economic dispatch problems with valve-point effects and multiple fuel options," *Energy*, vol. 116, pp. 637-649, 2016/12/01/ 2016.
4. M. H. Sulaiman and M. R. Mohamed, "Solving economic dispatch problems utilizing Cuckoo Search algorithm," in *2014 IEEE 8th International Power Engineering and Optimization Conference (PEOCO2014)*, 2014, pp. 89-93.
5. X.-S. Yang, S. S. Sadat Hosseini, and A. H. Gandomi, "Firefly Algorithm for solving non-convex economic dispatch problems with valve loading effect," *Applied Soft Computing*, vol. 12, pp. 1180-1186, 2012/03/01/ 2012.
6. M. H. Sulaiman, H. Daniyal, and M. W. Mustafa, "Modified Firefly Algorithm in solving economic dispatch problems with practical constraints," in *2012 IEEE International Conference on Power and Energy (PECon)*, 2012, pp. 157-161.
7. L. I. Wong, M. H. Sulaiman, M. R. Mohamed, and M. S. Hong, "Grey Wolf Optimizer for solving economic dispatch problems," in *2014 IEEE International Conference on Power and Energy (PECon)*, 2014, pp. 150-154.
8. S. Vijayaraj and R. K. Santhi, "Multi-area economic dispatch using flower pollination algorithm," in *2016 International Conference on Electrical, Electronics, and Optimization Techniques (ICEEOT)*, 2016, pp. 4355-4360.
9. F. Mohamed, M. Abdel-Nasser, K. Mahmoud, and S. Kamel, "Economic dispatch using stochastic whale optimization algorithm," in *2018 International Conference on Innovative Trends in Computer Engineering (ITCE)*, 2018, pp. 19-24.
10. H. Lotfi, A. Dadpour, and M. Samadi, "Solving economic dispatch in competitive power market using improved particle swarm optimization algorithm," in *2017 Conference on Electrical Power Distribution Networks Conference (EPDC)*, 2017, pp. 188-195.
11. K. Zare and T. G. Bolandi, "Modified iteration particle swarm optimization procedure for economic dispatch solving with non-smooth and non-convex fuel cost function," in *3rd IET International Conference on Clean Energy and Technology (CEAT) 2014*, 2014, pp. 1-6.
12. S. Chansareewittaya, "Hybrid BA/TS for economic dispatch considering the generator constraint," in *2017 International Conference on Digital Arts, Media and Technology (ICDAMT)*, 2017, pp. 115-119.
13. S. Mirjalili, "Moth-flame optimization algorithm: A novel nature-inspired heuristic paradigm," *Knowledge-Based Systems*, vol. 89, pp. 228-249, 2015/11/01/ 2015.
14. C. Po-Hung and C. Hong-Chan, "Large-scale economic dispatch by genetic algorithm," *IEEE Transactions on Power Systems*, vol. 10, pp. 1919-1926, 1995.
15. C.-C. Kuo, "A novel string structure for economic dispatch problems with practical constraints," *Energy Conversion and Management*, vol. 49, pp. 3571-3577, 2008/12/01/ 2008

Extraction and Filter Algorithm of Guidance Information for Full-strapdown Seeker on Rotation Missile

YongShan Liu[1], Li Song[1], and JingLong Li[1]

[1] *School of Aerospace Engineering, Beijing Institute of Technology, China*

Abstract. Strapdown seekers are superior to platform seekers for their simple structure, high reliability and light weight but cannot measure the line-of-sight angle rate information for the guidance of rotation missile directly. This paper aims at the engineering application of full-strapdown seekers on rotation missile problem. Firstly, a line-of-sight angle rate solution model is established. Based on the MATLAB, the extended Kalman filter (EKF) algorithm and unscented Kalman filter (UKF) algorithm are used to estimate the line-of-sight angle rate information of the full-strapdown seekers. The results show that using EKF filter and UKF filter both can obtain effective guidance information and the UKF's effect is better.

1 INTRODUCTION

Unlike traditional platform seeker, the full-strapdown seeker attaches all hardware to the projectile, removes the stable platform and the control mechanism. Which has the important significance on reducing the cost of seeker and improving the system reliability. Whereas the full-strapdown seeker cannot directly obtain the line-of-sight rate by its physical tracking loop and need digital calculation methods. Rotary missile system has the advantages of smaller size, lower cost, less control equipment, and can reduce the interference caused by the asymmetry of the projectile. However, the spin of the missile body will cause cross coupling between the pitch and yaw channels. Creagh MA et al. got the results through simulating which shows when the rotational speed of the projectile is high, the coupling caused by rolling becomes more serious, and due to the spin of the rotating projectile, the pitching channel and yawing channel of the rotating projectile will be cross-coupled [1].

As mentioned previously, the other estimation method of inertial line-of-sight rate is designed based on relative motion between missile and target, where many estimation algorithms are proposed according to the different filters adopted. Ehrich et al. put the additional rate compensation + differential network method to build single channel line-of-sight angle and then enter the guidance system through a low-pass filter [2]. Kim et al. [3] put forward a method to derive such correlation and finally obtained the inertial line-of-sight rate. During the deriving process, the body line-of-sight rate was calculated by a differential network. Bai Rui et al. [4] based on the typical launch conditions of small air-to-surface missile established an estimation algorithm and used the cubature Kalman filter for the rate estimation. Sun et al. [5] estimated the inertial line-of-sight rate of strapdown optical seeker using unscented Kalman filter (UKF) method, consider the estimation accuracy mainly depends on the body line-of-sight angle accuracy and gyro accuracy and simulate these two factors. Waldmann J [6] describes the modeling of an imaging seeker and the formulation of an extended Kalman filter for the estimation of line-of-sight rate from measurements of relative angular displacement between seeker gimbals and a low-cost strapdwn inertial unit. Li, J. J. [7] used the pursuit guidance law and proportion guidance law in order to get the line-of-sight rate and through amending "innovation" in the equations of Kalman filter, outlier-eliminating Kalman filtering algorithm is achieved.

In this paper, first based on the information coupling problem of rotation missile and the full-strapdown seeker problem, a decoupled model is established to extract the inertial the line-of-sight angle rate information from the observation data of full-strapdown seekers. And then by using the extended Kalman filter (EKF) algorithm and the unscented Kalman filter (UKF) algorithm, the real-time estimation of projectile guidance information is simulated and last verified the effectiveness of the algorithms.

2 LINE-OF-SIGHT ANGLE RATE MODEL

In this section, define four coordinate systems and the transformation between each coordinate system at the outset. Through introducing the missile–target distance and the approaching speed, the standard state equations and observation equations are presented. Getting the relation between the state equations and observation equation, the inertial line-of-sight elevation angle rate and the inertial line-of-sight azimuth angle rate are obtained.

2.1 Coordinate system definition

The following coordinate systems and angles will be used in this paper[8]:the inertial coordinate system $O_e x_e y_e z_e$, the body coordinate system $Ox_b y_b z_b$,the line-of-sight coordinate system $Ox_s y_s z_s$, and the body line-of-sight coordinate system $Ox_l y_l z_l$ .The inertial line-of-sight elevation angle q_γ ,the inertial line-of-sight azimuth angle q_λ ,the body line-of-sight elevation angle q_α ,the body line-of-sight azimuth angle q_β ,the pitch angle θ ,the yaw angle ψ ,the roll angle γ . The conversion of each coordinate system is shown in Fig.1.

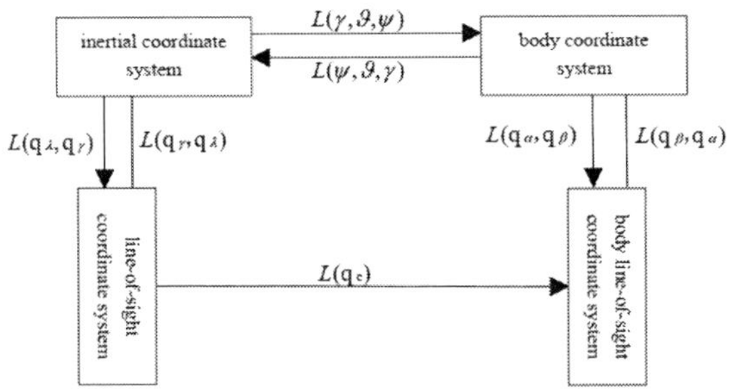

Fig 1. Conversion of each coordinate system.

2.2 Line-of-sight Information state equations

According to the kinematics and geometric relations of the rotation missile and target in space[9-10], the rotational angular velocity of the line-of-sight coordinate system $Ox_s y_s z_s$ relative to the ground coordinate system $O_e x_e y_e z_e$ is:

$$\omega_s = -\dot{q}_\lambda \sin q_\gamma \dot{i}_s + \dot{q}_\lambda \cos q_\gamma \dot{j}_s + \dot{q}_\gamma \dot{k}_s \tag{1}$$

$\dot{r}$ in Eq.(2) represents the change rate of the missile-target distant ,and in the line-of-sight coordinate system $Ox_s y_s z_s$ has the following relationship:

$$\begin{aligned} \dot{r}_x &= r\dot{q}_\lambda j - r\dot{q}_\lambda \cos q_\gamma k \\ \dot{r}_y &= r\dot{q}_\lambda \sin q_\gamma k - r\dot{q}_\lambda i \\ \dot{r}_z &= r\dot{q}_\lambda \cos q_\gamma i - r\dot{q}_\lambda \sin q_\gamma j \end{aligned} \tag{2}$$

Differentiate the missile-target relative speed V to obtain the missile-target relative acceleration a is:

$$a = \dot{V} = \begin{bmatrix} \ddot{r} - r\dot{q}_\gamma^2 - r\dot{q}_\lambda^2 \cos^2 q_\gamma \\ r\ddot{q}_\gamma + 2V\dot{q}_\gamma + r\dot{q}_\lambda^2 \sin q_\gamma \cos q_\gamma \\ 2r\dot{q}_\lambda \dot{q}_\gamma \sin q_\gamma - 2V\dot{q}_\lambda \cos q_\gamma - r\ddot{q}_\lambda \cos q_\gamma \end{bmatrix} \tag{3}$$

Assume that the target does not motorize, that is to say the missile-target relative acceleration $a = 0$, the Eq. (3) can be obtained:

$$\begin{cases} \ddot{r} = r(\dot{q}_\gamma^2 + \dot{q}_\lambda^2 \cos^2 q_\gamma) \\ \ddot{q}_\gamma = \dfrac{-2V\dot{q}_\gamma}{r} - \dot{q}_\lambda^2 \sin q_\gamma \cos q_\gamma \\ \ddot{q}_\lambda = 2\dot{q}_\lambda \dot{q}_\gamma \tan q_\gamma - \dfrac{2V\dot{q}_\lambda}{r} \end{cases} \tag{4}$$

The state vector of the system can be augmented as $x = [x_1 \quad x_2 \quad x_3 \quad x_4]^T = [q_\lambda \quad \dot{q}_\lambda \quad q_\gamma \quad \dot{q}_\gamma]^T$ .Thus, the line-of-sight information state equations are derived by combining Eq. (4) as follows:

$$\begin{cases} \dot{x}_1 = x_2 \\ \dot{x}_2 = -\dfrac{2V}{r} x_2 + 2x_2 x_4 \tan x_3 \\ \dot{x}_3 = x_4 \\ \dot{x}_4 = -x_2^2 \sin x_3 \cos x_3 - \dfrac{2V}{r} x_4 \end{cases} \tag{5}$$

2.3 Observation Equations

Assume the position of the target in the line-of-sight coordinate system and the body line-of-sight coordinate system are both $[R \quad 0 \quad 0]$ .And the coordinates of the target and the missile in the inertial coordinate system are $[x_{eT} \quad y_{eT} \quad z_{eT}]^T$ and $[x_{eD} \quad y_{eD} \quad z_{eD}]^T$. According to the conversion relationship between the line-of-sight coordinate system $Ox_s y_s z_s$ and the inertial coordinate system $O_e x_e y_e z_e$, the relative equations can be written as:

$$\begin{bmatrix} x_{TD} \\ y_{TD} \\ z_{TD} \end{bmatrix} = \begin{bmatrix} x_{eT} \\ y_{eT} \\ z_{eT} \end{bmatrix} - \begin{bmatrix} x_{eD} \\ y_{eD} \\ z_{eD} \end{bmatrix} = \begin{bmatrix} R\cos q_\gamma \cos q_\lambda \\ R\sin q_\gamma \\ -R\cos q_\gamma \sin q_\lambda \end{bmatrix} \tag{6}$$

The geometric relative equations are Eq. (7).

$$\begin{cases} q_\gamma = \arctan\left(\dfrac{y_{TD}}{\sqrt{x_{TD}^2 + z_{TD}^2}}\right) \\ q_\lambda = \arctan\left(-\dfrac{z_{TD}}{x_{TD}}\right) \end{cases} \tag{7}$$

In the similar way, define the coordinates of the target and missile in the body coordinate system are $[x_{bT} \quad y_{bT} \quad z_{bT}]^T$ and $[x_{bD} \quad y_{bD} \quad z_{bD}]^T$. The conversion relationship between the body coordinate system $Ox_b y_b z_b$ and the body line-of-sight coordinate system $Ox_l y_l z_l$:

$$\begin{bmatrix} \tilde{x}_{TD} \\ \tilde{y}_{TD} \\ \tilde{z}_{TD} \end{bmatrix} = \begin{bmatrix} x_{bT} \\ y_{bT} \\ z_{bT} \end{bmatrix} - \begin{bmatrix} x_{bD} \\ y_{bD} \\ z_{bD} \end{bmatrix} = \begin{bmatrix} R\cos q_\alpha \cos q_\beta \\ R\sin q_\alpha \\ -R\cos q_\alpha \sin q_\beta \end{bmatrix} \quad (8)$$

The geometric relations are:

$$\begin{cases} q_\alpha = \arctan\left(\dfrac{\tilde{y}_{TD}}{\sqrt{\tilde{x}_{TD}^2 + \tilde{z}_{TD}^2}}\right) \\ q_\beta = \arctan\left(-\dfrac{\tilde{z}_{TD}}{\tilde{x}_{TD}}\right) \end{cases} \quad (9)$$

R_{ij} is the element of the conversion matrix C_e^b in the i th row and j th column which is the coordinate transformation matrix from the ground coordinate system to the body coordinate system,. The conversion C_e^b matrix:

$$C_e^b = \begin{bmatrix} \cos\vartheta\cos\psi & \sin\vartheta & -\cos\vartheta\sin\psi \\ -\sin\vartheta\cos\psi\cos\gamma+\sin\psi\sin\gamma & \cos\vartheta\cos\gamma & \sin\vartheta\sin\psi\cos\gamma+\cos\psi\sin\gamma \\ \sin\vartheta\cos\psi\sin\gamma+\sin\psi\cos\gamma & -\cos\vartheta\sin\gamma & -\sin\vartheta\sin\psi\sin\gamma+\cos\psi\cos\gamma \end{bmatrix} \quad (10)$$

Combine the Eq. (6), Eq. (8) and Eq. (10) can get:

$$\begin{bmatrix} R\cos q_\alpha \cos q_\beta \\ R\sin q_\alpha \\ -R\cos q_\alpha \sin q_\beta \end{bmatrix} = \begin{bmatrix} R_{11} & R_{12} & R_{13} \\ R_{21} & R_{22} & R_{23} \\ R_{31} & R_{32} & R_{33} \end{bmatrix} \begin{bmatrix} R\cos q_\gamma \cos q_\lambda \\ R\sin q_\gamma \\ -R\cos q_\gamma \sin q_\lambda \end{bmatrix} \quad (11)$$

Angle information provided by full-strapdown seeker includes the body line-of-sight angle q_α and q_β, which are considered as the observation variables. Under this circumstances, defining the line-of-sight observation vector $y = \left[q_\alpha \quad q_\beta\right]^T = \left[y_1 \quad y_2\right]^T$. The line-of-sight observation equations of the full-strapdown seeker can be written as:

$$\begin{cases} y_1 = \arcsin(R_{21}\cos x_3 \cos x_1 + R_{22}\sin x_3 - R_{23}\cos x_3 \sin x_1) \\ y_2 = -\arctan(\dfrac{R_{31}\cos x_1 + R_{32}\tan x_3 - R_{33}\sin x_1}{R_{11}\cos x_1 + R_{12}\tan x_3 - R_{13}\sin x_1}) \end{cases} \quad (12)$$

So far, the line-of-sight angle information state equations and the observation equations of the of the full-strapdown seeker on the rotating missile are established. From Eq. (5) and Eq. (12), it can be seen that there is strong nonlinearity existing both in the line-of-sight state equations and observation equations.

3 Rotation Missile Guidance Information Estimation and Simulation

In this paper, the Extended Kalman Filter (EKF) and Unscented Kalman Filter (UKF) algorithm are simulated and verified under both the navigation system measurement data and the stapling target coordinates have errors. In the inertial coordinate system, the actual position of the target is (0m, 2.5m, 0m), the coordinate of stapling target is (100m, 52.5m, 60m). That is, the stapling target measurement error is (100m, 50m, 60m). The initial position of the missile in the ground coordinate system is (-7900m, 2050m, 60m).

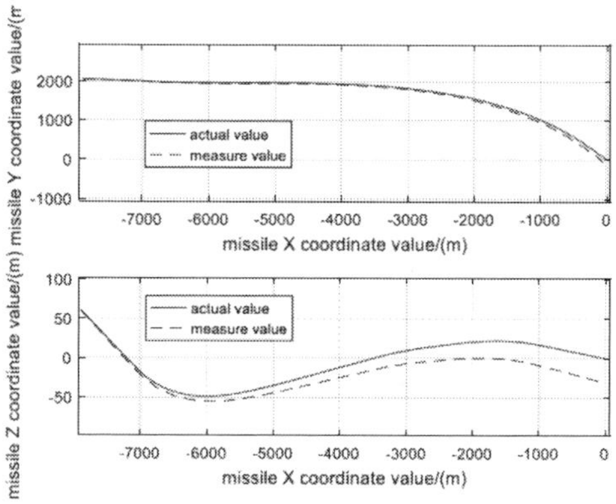

Fig 2. Missile X, Y, Z location.

It can be seen from the Fig. 2 that the measurement errors between the actual position and measurement position of the rotation missile in the inertial navigation system gradually increases with time. Therefore, directly using the data measured by the inertial navigation system to guide would miss targets.

In the process of the rotation missile flying to the target, the body line-of-sight elevation angle q_α and the body line-of-sight azimuth angle q_β are shown in Fig. 3.

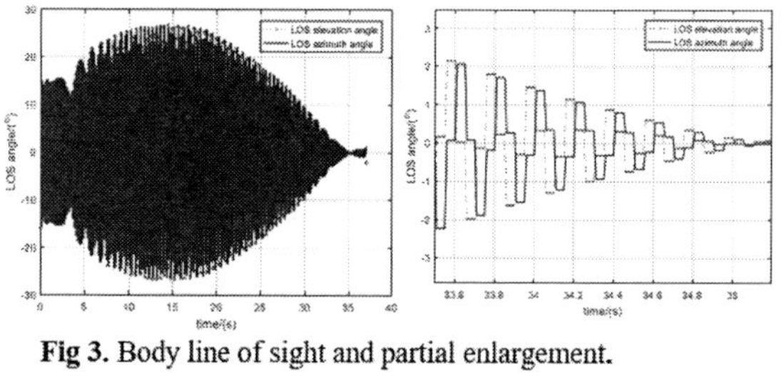

Fig 3. Body line of sight and partial enlargement.

Since the rotating missile will generate the pitch and yaw directions coupling during the movement, the body line-of-sight elevation angle q_α and the body line-of-sight azimuth angle q_β in Fig. 3 are same size and 90 degrees phase difference. Due to the low repetition rate of the laser pulse code, the measured signals of the body line-of-sight elevation angle and the body line-of-sight azimuth angle are not perfect as sine functions.

3.1 Estimation of Guidance Information Based on EKF and UKF Algorithms

The stochastic nonlinear discrete formula for the extended Kalman (EKF) filtering algorithm [11] is:

$$X_k = \Phi_{k,k-1}X_{k-1} + \Gamma(\hat{X}_{k-1,k-1})W_{k-1} \quad (13)$$

$$Z_k = H_k X_k + V_k \quad (14)$$

In the Eq.(13) and Eq.(14), W is the process noise matrix, V is the observation noise matrix.

The line-of-sight information of rotating missile has strong nonlinear characteristics. When applying extended Kalman filter, the Jacobian matrix is needed to obtain the deviation based approximate linear equations of the state equations.

$$J_{k,k-1} = \frac{\partial f}{\partial X_{k-1}^{*}} = \left.\frac{\partial f(X_{k-1},k-1)}{\partial X_{k-1}}\right|_{X_{k-1}=X_{k-1}^{n}} \quad (15)$$

Discretize the Jacobian matrix:

$$\Phi_{k,k-1} = I + J_{k,k-1}\cdot\Delta t \quad (16)$$

Unscented kalman filter (UKF) algorithm firstly needs to designs a series of Sigma points. Then calculating the result of $f(\cdot)$ with the set sigma points, and $(\bar{Z},Pz)$ based on χ_i.

$$\begin{cases} \xi_{k-1}^{(0)} = \hat{X}_{k-1}^{a} \\ \xi_{k-1}^{(i)} = \hat{X}_{k-1}^{a} + (\sqrt{(n+\lambda)P_{k-1}^{a}})_i, i = 1,2,\dots,n \\ \xi_{k-1}^{(i)} = \hat{X}_{k-1}^{a} - (\sqrt{(n+\lambda)P_{k-1}^{a}})_i, i = n+1,n+2,\dots,2n \end{cases} \quad (17)$$

Propagate $\chi_{k,k-1}$ through nonlinear observation function $H(\cdot)+r_k$. From $\chi_{k,k-1}$ can get the output prediction $\hat{Z}_{k,k-1}$, as well as the self-covariance matrix $P_{\tilde{Z}_k}$ and the mutual covariance matrix $P_{X_kZ_k}$:

$$\chi_{k,k-1} = H_k(\xi_{i,k-1},v_k)+r_k, i = 0,1,\dots,2n \quad (18)$$

$$\hat{Z}_{k,k-1} = \sum_{i=0}^{2n}\omega_i^{(m)}\chi_{i,(k,k-1)} = \sum_{i=0}^{2n}\omega_i^{(m)}H_k(\xi_{i,k-1},v_k)+r_k \quad (19)$$

$$P_{\tilde{Z}_k} = \sum_{i=0}^{2n}\omega_i^{(c)}(\chi_{i,(k,k-1)}-\hat{Z}_{k,k-1})(\chi_{i,(k,k-1)}-\hat{Z}_{k,k-1})^T + R_k \quad (20)$$

$$P_{X_kZ_k} = \sum_{i=0}^{2n}\omega_i^{(c)}(\xi_{i,(k,k-1)}-\hat{X}_{k,k-1}^{a})(\chi_{i,(k,k-1)}-\hat{Z}_{k,k-1})^T \quad (21)$$

So far, we can use the EKF filter algorithm and UKF filter algorithm to extract the line-of-sight angle race of the full-strapdown laser seeker on the rotation missile as the guidance information [12]. It is estimated and simulated as shown in Fig.4 and Fig.5.

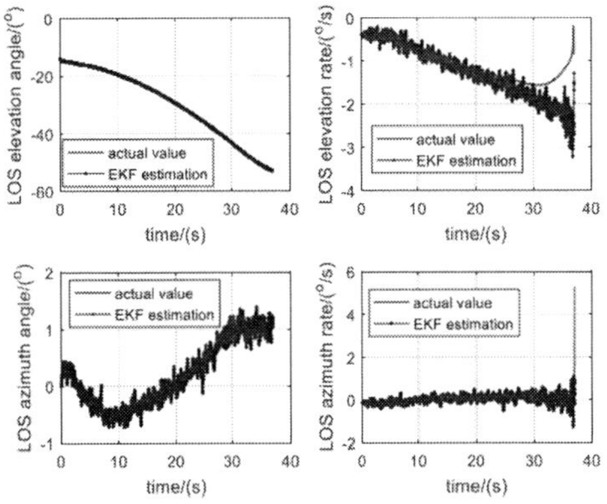

Fig4. EKF filtering line-of-sight and line-of-sight rate estimation.

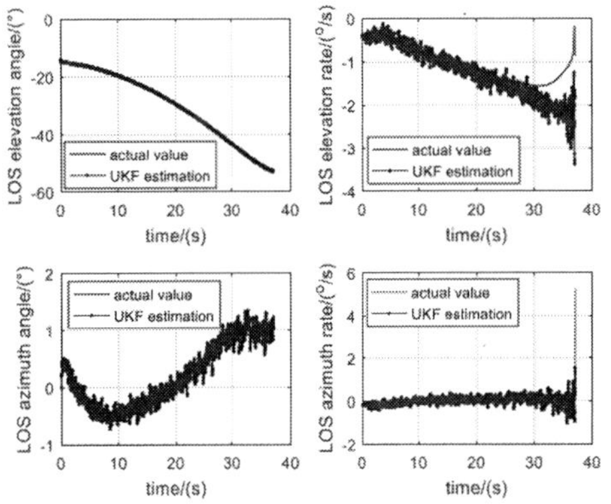

Fig 5. UKF filtering line-of-sight and line-of-sight rate estimation.

3.2 Computational error analysis

Calculate the EKF and UKF errors between the filter value and the true value on the line-of-sight elevation angle, the line-of-sight elevation race, the line-of –sight azimuth angle and the line-of-sight azimuth race.

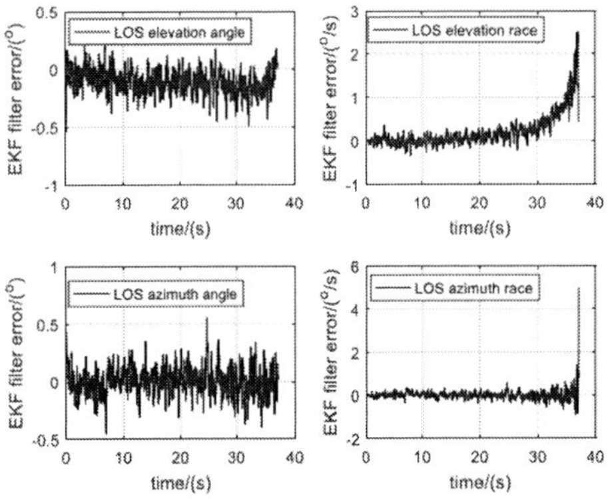

Fig 6. Errors between EKF filter and true value.

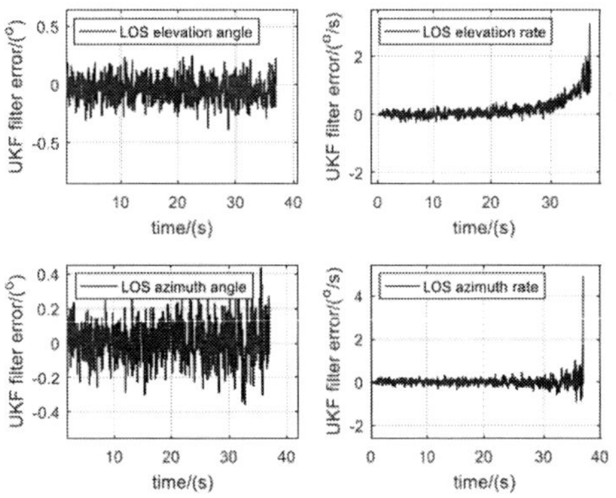

Fig 7. Errors between UKF filter and true value.

It can be seen from the Fig.6 and Fig.7 that if the non-linearity is not strong, the results of the EKF filter and the UKF filter are similar. But in the case of high non-linearity such as the line-of-sight elevation race, the UKF filtering is better than the EKF filtering.

Calculate the EKF and UKF mean error between the filter value and the true value of the on the line-of-sight angle and the line-of-sight rate gets the results shown in Table 1:

Table 1. Mean error of EKF and UKF filtering information.

Mean error	EKF filter	UKF filter
LOS elevation angle	0. 234426	0. 088877
LOS elevation race	0. 347482	0. 133783
LOS azimuth angle	0. 212946	0. 092201
LOS azimuth race	0. 318609	0. 186944

Calculate the root-mean-square error ((RMSE) of the EKF filter and UKF filter can obtain Table.2.

Table 2. RMSE of EKF and UKF filtering information.

RMSE	EKF filter	UKF filter
LOS elevation angle	0. 115062	0. 095683
LOS elevation race	0. 354065	0. 194481
LOS azimuth angle	0. 104441	0. 085619
LOS azimuth race	0. 204107	0. 142643

4 CONCLUSION

This paper aims at the engineering application problem of the full-strapdown laser seeker on a rotation missile. First establishing a line-of-sight rate information extraction model. Based on the model, using EKF filter algorithm and UKF filter algorithm to filter and estimate the line-of-sight angle and rate information of the full-strapdown laser seeker. And verifying the filtering algorithm by MATLAB simulation. The results show that UKF filter is better than EKF filter and can be used to extract the guidance information of full-strapdown seeker for rotation missile more accurate.

In this paper, because of time constraints we just targeted research the full-strapdown on rotation missile, other types of seeker algorithms, such as infrared radar, can be further studied. And if more time permits, the control system of rotating projectile can be further studied to realize projectile body integration.

References

1. Creagh MA, Mee D J. Attitude guidance for spinning vehicles with independent pitch and yaw control[J]. Journal of guidance, control, and dynamics, 33(3),(2010)
2. Ehrich R D, Vergez P. Strapdown seeker technology for the terminal guidance of tactical weapons[C],AGARD Guidance and Control Aspects of Tactical Air launched Missiles 15p,1980, 81(10): 7-15.
3. Kim, D., Ryoo, C. K., Kim, Y., and Kim, J., Guidance and Control for Missile with a Strapdown Seeker, Proceedings of the 11th International Conference on Control, Automation and Systems, Vol. 86, IEEE Publ.,Piscataway, NJ, 2011, pp. 969–972.
4. Bai Rui,Xia Qunli, Zhang Daochi, Technology of line-of-sight rate estimation using STCKF for strapdown seeker, Infrared and Laser Engineering,Vol.46 No.11,(2017)
5. Sun T, Chu H, Zhang B, et al. Line-of-sight rate estimation based on UKF for strapdown seeker [J].Mathematical Problems in Engineering, 2015, 2015(1): 1-14.
6. Waldmann J.Line-of-sight rate estimation and linearizing control of an imaging seeker in a tactical missile guided by proportional navigation. IEEE Transactions on control systems technology.10, (2002)
7. Li, J. J., "Research of LOS Rate Estimation Method for Strapdown Imaging Seeker," M.S. Thesis, Harbin Inst. of Technology, Harbin, Heilongjiang Province, PRC, (2008).
8. Qian Xingfang, Lin Ruixiong, Zhao Yanan, Missile flight mechanics (in Chinese), Beijing Institute of Technology Press, China, (2013).
9. James, M. M., "Line of Sight Rate Estimation for Guided Projectiles with Strapdown Seeker," AIAA GNCC, (2015).
10. Joongsup Yun, Chang-Kyung Ryoo, Taek-Lyul Song, Strapdown sensors and seeker based guidance filter design, ICCAS, 468-472, (2008).
11. Julier, S. J., and Uhlmann, J. K., "Unscented Filtering and Nonlinear Estimation," Proceedings of the IEEE, No. 3, (2004)
12. Luo Yufeng, Liu Yong, The simulation research on the strapdown inertial navigation algorithm based on the trajectory generator (in Chinese), Journal of Henan Polytechnic University (Natural Science), 06,(2015)

Toward Wearable EEG-based Alertness Detection System Using SVM with Optimal Minimum Channels

Mihong Yang , Huiyan Li , Xiaozhou Sun , Li Yang , Hailong Duan , Yanqiu Che and Chunxiao Han

Tianjin Key Laboratory of Information Sensing & Intelligent Control, School of Automation and Electrical Engineering, Tianjin University of Technology and Education, Tianjin, 300222, P. R. China

Abstract. Alertness is the state of attention by high sensory awareness. A lack of alertness is one of the main reasons of serious accidents. Traffic accidents caused by driver's drowsy driving have a high fatality rate. This paper presents an EEG-based alertness detection system. In order to ensure the convenience and long-term wearing comfort of EEG recordings, the wearable electrode cap will be the principal choice in the future, and the selection of channels will be limited. We first built a 3-D simulated driving platform using Unity3D. Then, we perform an experiment with driving drift task. EEG signals are recorded form frontal and occipital regions. We select data segments using the driving reaction time, classify the state of alertness with a support vector machine (SVM), and select the optimal combination of channels with minimum number of channels. Our results demonstrate that alertness can be classified efficiently with one channel (PO6) at accuracy of 93.52%, with two channels (FP1+PO6) at 95.85% and with three channels (FP1+PO6+PO5 and FP1+PO6+POZ) at 96.11%.

1 Introduction

The issue of a decrease in the degree of alertness at work has received increasing attention due to the gradual increase of living pressure and work intensity [1]. Alertness refers to a person's ability to maintain his focus of attention on performing an operational task for prolonged periods of time. For example, the alertness level of a driver refers to the ability of the driver to perceive and understand the status of the car or the traffic conditions in a timely manner while driving, and to make the correct operation [2]. Staying alertness above a certain level is very important in some cases in our daily lives [1]. In general, drowsy driving will reduce thinking ability of driver, and also is a risk factor for traffic crashes and deaths [3, 4]. If the driver feels fatigue, it will seriously threaten the life safety of driver and even others. In the United States, as reported by the National Traffic Safety Administration, 60,000 traffic accidents take place due to pilot fatigue driving every year [5]. Traffic accidents in China in recent years, the burst frequency obviously increase, and the driver's low- alertness is one of the main reasons of serious accidents. Therefore, detecting alertness level of the driver is of great significance to improve the driving features and the traffic safety situation.

A number of factors could contribute to decrease alertness level of a driver, such as irritability, lack of sleep, or long driving the car. Many methods have been proposed to detect changes of the level of alertness. These methods can be divided into three main approaches until now. The first method is to record the score of Karolinska Sleepiness Scale (KSS) [6] or Stanford Sleepiness Scale (SSS) [7] to detect the state of alertness, and the method is usually an auxiliary method. The second method is to analyze the behaviors of driver including blink, yawn and sagging posture with video technology [4, 8, 9], which has been widely used to detect the level of alertness. But, the video technology also has some disadvantages, for example, the collection of video in night is difficult and it also limit the driver's movement [10]. The third method focuses on detecting physiological changes of drivers, such as electroencephalographic (EEG) activities and electrooculographic (EOG) activities [11-17]. Many researchers choose the EEG-based method to detect the state of alertness due to EEG signals have advantages for making accurate judgments of driver's alertness levels in time, and we use the EEG-based method in this study.

So far, many researchers have used EEG signals to detect alertness. Subasi [18] used C3-A2 channels EEG signals to assess a driver's alertness and measure the potential applicability of the artificial neural network (ANN). Mervyn et al. [19] used 32 EEG channels to classify the data using support vector machine (SVM) with 99.3% accuracy. However, in order to ensure the convenience and long-term wearing comfort of EEG experiments, the wearable electrode cap will be the principal choice, and the positions of channels on the head will be limited. Thus, we chose 10-channel EEG signals located at the frontal area and the occipital area to detect the state of subject's alertness. We developed a

https://doi.org/10.1051/matecconf/201821403009

simulated driving platform for experiments and used an efficient classifier-SVM to classification. Finally, we obtained relatively ideal classification results through the optimal combination of channels with the smallest number of channels. The variance and the peak as the feature vector were used for classification of alertness, and the method of feature extraction we used in this study is simple. So, we used a lowpass filter with a cut-off frequency of 40 Hz to remove the high frequency noise and to get better characters from EEG data [20].

This paper is organized as follows. In Section 2, we describe the experiments setup, including the information of the subjects, Unity3D-based driving platform and data collection. Section 3 introduces the analysis methods, including the data segmentation and feature extraction. Section 4 presents the classification results. Finally, we conclude this paper in Section 5.

2 Experiments setup

2.1 Subjects

Simulation driving experiments were performed in one volunteer, Male, at age of 36. Statistical reports [21] indicated that the drowsiest time also occurs during the early afternoon hours, except from late night to early morning. And drowsiness often occurs within one hour of continuous driving during these periods. And the time for starting the simulated driving experiments is the afternoon hours after lunch because the subject easily become drowsy within an hour of continuous driving. The subject was told of the driving task, and then had a 15-minute practice [20] to keep the car at the middle of the third lane of the road by controlling with the steering wheel. The subject has signed an informed written consent with the adequate understanding of the procedure of the experiment.

2.2 Unity3D-based driving platform

In this study, we developed a simulation driving platform for highway scenes using the Unity3D software. First, we created some models of objects including trees, hills, roads, and cars for the scene and set up the relative parameters of the speed, form, and direction of the car. Then, we built a complete simulated driving scene of full functionality with these virtual objects and the aid of the C# programming language. Fig. 1 shows the Unity3D-based driving scene including four lanes from left to right. The driving scene appears cyclically as the subject is driving the car at an adjustable velocity on the road. During the driving, the car is randomly drifted away from the center of the third lane of the road, according the consequences of a real road surface. The simulated driving software communicated with a Neuroscan 64-channel physiological measuring system, where the EEG signals and event information were continuously recorded. Fig. 2 shows the scene of experiment.

Figure 1. Unity3d-based driving scene used in our experiments.

Figure 2. The scene of experiment

2.3 Data collection

During each experiment, 10 EEG channels and the event information containing event type and time were recorded by the NeuroScan NuAmpsExpress system. The contact impedance between scalp and EEG electrodes was reduced below 5 kΩ before data acquisition. Each subject's EEG data were recorded for 1 hour at a sampling rate of 1000 Hz. Because of limitations of the wearable electrode cap, we chose 10 EEG channels located at FP1, FP2, PO5, PO3, POZ, PO4, PO6, O1, OZ and O2 with reference electrode placed at the vertex (CZ), as shown in Fig. 3.

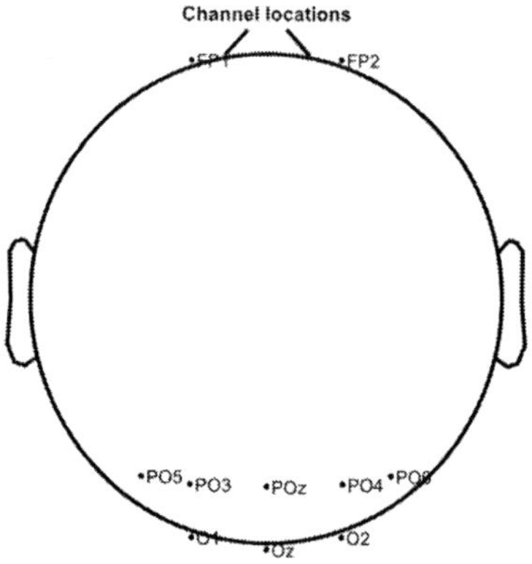

Figure 3. Electrode names and positions on the head

3 Analysis methods

The flowchart of data analysis for alertness classification with the optimal combination of channels is shown in Fig. 4. For each subject, after collecting 10-channel EEG signals in 1 hour simulation driving experiment, the EEG data was preprocessed using a lowpass filter with a cut-off frequency of 40 Hz to remove the high frequency noise and to get better characters. Then, we selected the segments by judging the reaction time. For the segments including drowsy segments and alert segments, we extracted the variance and peak as features separately. The efficient classifier SVM was used to determine the classification of these feature vectors: 1 for drowsiness and -1 for alertness. Finally, we selected the minimum and the best channels from the 10 EEG channels.

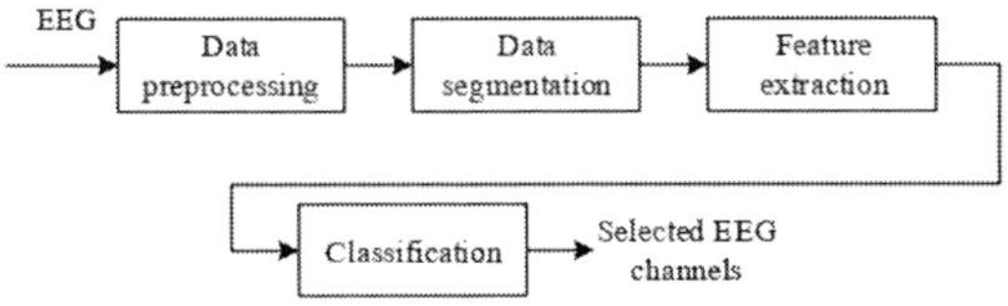

Figure 4. Flowchart for processing the EEG signals

3.1 Data segmentation

Our object is to achieve the classification of alertness with the best combination and the smallest number of channels. Thus, we first need to prepare the dataset including drowsy segments and alert segments from 10-channel EEG signals. To select segments as training set and testing set, and to classify the state of the subject's alertness, we defined a subject's driving reaction time as the difference between the time of car's drift and the time of subject's correction. When the subject was drowsy, the value of driving reaction time increased, as shown in Fig. 5. We calculated the moving averages of subject's driving reaction time using a 60-point window with 30-point overlap, and selected averages of more than 2.5s as drowsy segments, as shown in Fig. 6 (left), and selected averages of less than 1.5s as alert segments, as shown in Fig. 6 (right). Then, we intercepted the half of drowsy segments and alert segments as training set and testing set, respectively. Each data set was intercepted using 5000-point window with 3000-point overlap as samples. Finally, 2 drowsy segments and 2 alert segments were selected and each segment contained about 200 samples (400s).

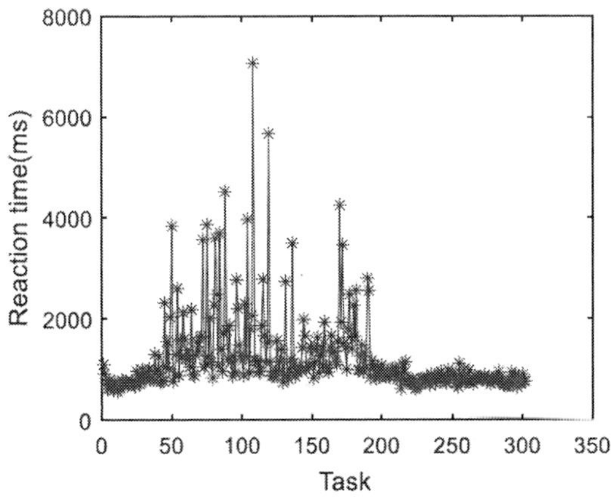

Figure 5. Subject's reaction time for drift

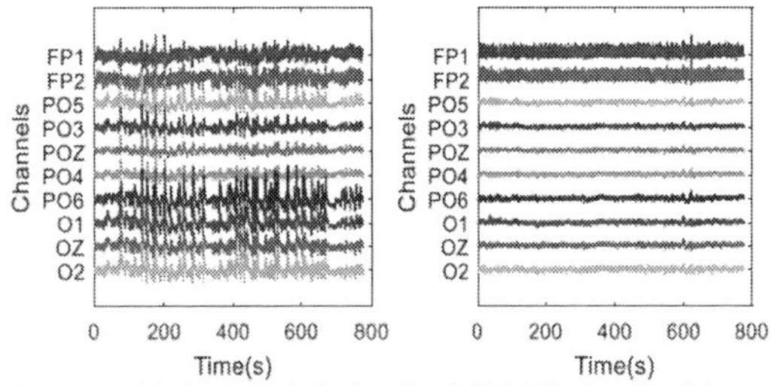

Figure 6. 10-channel EEG signals: (left) EEG signals of drowsy state and (right) EEG signals of alert state

3.2 Feature Extraction

In this study, features we used in classification include the variance and the peak of each sample. And the variance and the peak as the feature vector were used for classification of alertness. Each drowsy segment or alert segment contained 60 drifts, about 800s. With a 5s window with 3s overlap, the training set contained about 400 samples. Fig. 7 showed features computed from EEG signals of drowsy state and alert state.

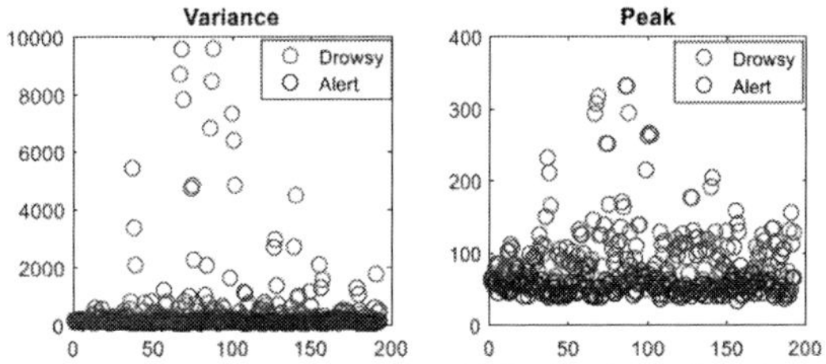

Figure 7. Features computed from EEG signals. (left) variance (right) peak values. Red circles represent drowsy states, and blue circles represent alert states.

4 Results and Discussion

In this study, we used the driving reaction time to select data segments as training set and testing set, and used the variance and peak as the feature vector to compute features of 10-channel EEG signal by the sliding window technique. And the sliding window length was 5s with 3s overlap. The SVM was used to determine the classification in 5s epochs of these feature vectors: 1 for drowsiness and -1 for alertness. When the number of channels was changed, the results of classification were different, as shown in Table 1. It is observed that the effect of classification by using all channels is better than using single channel, and using part of channels is better than using all channels. Thus, in order to find the optimal combination of channels, we computed the classification accuracy of different combinations to evaluate the results [22].

Firstly, we used all channels and a single channel to determine the classification respectively, and found that accuracy of all channels was 95.34%, and it was higher than accuracies of any single channel. PO6 reached the best performance of a single channel with accuracy of 93.52%. Then we used two channels to compute accuracies, and there were 45 combinations out of 10 channels. In these combinations of channels, we found that the combination of channels of frontal area and occipital area could improve the accuracy, and the top 11 results from high to low were shown in Table 1. And it was consistent with Ahmed Al-Ani [23]. For the drowsy state, the selected channels were from with the parietal/occipital areas, including the frontal area. According to the results of classification of two channels, we chose 2 combinations with higher accuracy to further research, and the accuracy of combination of FP1 and PO6 was 95.85%, and the accuracy of combination of FP1 and PO5 was 95.34%. Then, we added other channels to these 2 combinations respectively, and found that the combination of FP1, PO5 and PO6 and FP1, POZ and PO6 got the highest accuracy of 96.11%. We also chose 4 to 9 channels from 10-channel as combinations and found that accuracies of these combinations were rather lower than previous combinations. And the accuracy did not increase with the increase of the number of channels. In order to select the optimal combination from the combination of FP1, PO5 and PO6 and the combination of FP1, POZ and PO6, we increased channels to two combinations respectively, as shown in Table 1. In consideration of the symmetry of channels and the development of the wearable electrode cap, thus, we suggested the optimal combination with the smallest number of channels to be FP1, PO5 and PO6.

Table 1. Accuracy (%) using different combinations of channels with SVM classifier.

channels	SVM	channels	SVM
All	95.34	FP1+PO6	95.85
FP1	88.34	FP1+PO5	95.34
FP2	82.38	FP1+OZ	95.08
PO5	91.71	FP2+PO6	94.82
PO3	90.41	FP1+O1	94.56
POZ	90.67	FP1+POZ	94.30
PO4	89.12	FP1+O2	94.30
PO6	**93.52**	FP2+O1	93.52
O1	92.23	PO5+PO6	93.52
OZ	89.38	FP1+PO3	93.26
O2	89.90	FP2+PO5	93.26
FP1+PO6+PO5	96.11	FP1+PO5+PO6+FP2	96.11
FP1+PO6+POZ	96.11	FP1+PO5+PO6+PO3	96.11
FP1+ PO6+FP2	95.60	FP1+PO5+PO6+POZ	95.85
FP1+PO6+PO3	95.60	FP1+PO5+PO6+PO4	95.34
FP1+PO6+O1	95.60	FP1+PO5+PO6+O1	96.11
FP1+PO6+PO4	95.85	FP1+PO5+PO6+OZ	95.85
FP1+PO6+O2	95.85	FP1+PO5+PO6+O2	95.34
FP1+ PO6+OZ	94.82	FP1+POZ+PO6+FP2	96.11
FP1+PO5+OZ	95.34	FP1+POZ+PO6+PO3	95.34
FP1+PO5+FP2	94.82	FP1+POZ+PO6+PO4	95.85
FP1+PO5+O1	94.56	FP1+POZ+PO6+O1	96.11
FP1+PO5+O2	94.56	FP1+POZ+PO6+OZ	95.34
FP1+PO5+POZ	94.30	FP1+POZ+PO6+O2	95.34
FP1+PO5+PO3	93.52		
FP1+PO5+PO4	93.52		

5 Conclusions

In this study, we have developed a simulation driving platform for highway scenes using the Unity3D software, and have designed an experiment that the car was randomly drifted away from the center of the third lane of the road. In order to follow the development of the wearable electrode cap, we have chosen 10 EEG channels located at the frontal area and the occipital area. We selected the data into drowsy segment or alert segment by the driving reaction time. The variance and the peak have been extracted as feature vectors. We have obtained the classification of the subject's alertness using SVM with the optimal combination of minimum channels. Our results demonstrated that the optimal combination with the smallest number of channels is FP1, PO5 and PO6, with accuracy of 96.11%. In the further work, we will select more features as feature vectors to process the data and classify the alertness. And according to the optimal combination with the smallest number of channels, we will further analyse the relationship between EEG data and driving behaviour. The methods we used in this study are simple and efficient, and might be used to classify the state of the driver's alertness for a real time wearable alertness-monitoring system.

References

1. S. J. Weston, J. J. Jackson, J. Res. Pers, **73**, 27 (2018)
2. E. Aidman, C. Chadunow, K. Johnson, J. Reece, Accident Anal. Prev, **81**, 8 (2015)
3. National Highway Traffic Safety Administration. Traffic safety facts crash stats: drowsy driving. Available at: https://crashstats.nhtsa.dot.gov/Api/Public/ViewPublication/811449.
4. KU Anjali, AK Thampi, A Vijayaraman, *International Conference on Circuit*, 1 (2016)
5. Nat. *Highway Traffic Safety Admin*. U. S. Dept. Transp. Tech. Rep. DOTHS 810818 (2006)
6. Otmani S, Pebavle T, Roge J, Physiol. Behav, **84**, 715 (2005)
7. Hoddes E, Dement W, Zarcone V, Psychophysiology, **9**, 150 (1972)
8. Ji Q, Zhu Z, Lan P, IEEE T. Veh. Technol, **53**, 1052 (2004)
9. BG Lee, SJ Jung,WY Chung, et al, *IET Commun*, **5**, 2461 (2011)
10. Z. Zhang, J. Zhang, Journal of Control Theory and Application, **8**, 181 (2010)
11. S. Makeig, T. P. Jung, Cognitive Brain RE S, **4**, 15 (1996)
12. K. Van Orden, W. Limbert, S. Makeig, T.-P. Jung, Hum. Factors, **43**, 111 (2001)
13. J. Qiang, Z. Zhiwei, P. Lan, IEEE Trans. Veh. Technol, **53**, 1052 (2004)
14. S. Makeig M. Inlow, Electroencephalogr Clin. Neurophysiol, **86**, 23 (1993)
15. T. L. Morris J. C. Miller, Biol. P Sychol., **42**, 343 (1996)
16. T.-P. Jung, S. Makeig, M. Stensmo, T. J. Sejnowski, IEEE Trans. Biomed. Eng., **44**, 60 (1997)
17. S. Roberts, I. Rezek, R. Everson, H. Stone, S. Wilson, C. Alford, IEE P–SCI Meas. Tech, **147**, 333 (2000)
18. A. Subasi, Expert Sys. Appl., **28**, 701 (2005)
19. M. V. M. Yeo, X. Li, K. Shen, E. P. V. Wilder-smith, J. Safety Science, **47**, 115 (2009)
20. Lin C T, Wu R C, Jung T P, et al. Eurasip J. Appl. Sig. P., **19**, 1 (2005)
21. H. Ueno, M. Kaneda, M. Tsukino, *Vehicle Navigation and Information Systems Conference (VNIS '94)*, Yokohama, Japan: August–September (1994)
22. Y. Yang, I. Bloch, S. Chevallier, J. Wiart, Cogn. Comput, **8**, 505 (2016)
23. A. Ani, I. Koprinska, G. Naik, Expert Sys. Appl., **83**, 273 (2017)

Plant Layout and Simulation Roaming System based on Virtual Reality Technology

Liang Dong[1], Zhifeng Wang[1]

[1] *Shanghai Second Polytechnic University, Institute of intelligent manufacturing and control engineering, Shanghai 201209, china*

Abstract. The layout and roaming of machinery and equipment in the workshop is an important development direction of intelligent manufacturing. In this paper, the general process of developing an intelligent factory system are put forward. The 3D models of various kinds of mechanical equipment are established. The textures and maps of the model are processed. The virtual simulation engine is used to establish the workshop map. The roaming work of the virtual workshop is developed based on the map. The space layout of the mechanical equipment and the machine tool simulation cutting system are developed. Comparing with the traditional simulation software, this system takes shorter period of development and achieves better visualization. And combined with the virtual reality technology makes this system more creative.

1 Introduction

Virtual reality (VR) is a kind of computer simulation system that can create the virtual world and let people experience it. It uses computer to generate simulation environment. It is a system simulation with multi-source information fusion, interactive three-dimensional dynamic view and entity behaviour [1-3]. Recent years, numerous research achievements of introducing VR technology into the field of intelligent plant construction have been made. Yu Qiang and others studied the layout of factory processes in three-dimensional interactive environment [4]. Wen Huaixing and others designed a virtual assembly system of parallel machine tools based on Virtools [5]. Liang Hongbao discussed three-dimensional scene modelling technology and 3D scene optimization technology [6]. Zhu Wenhua developed the engineering training system of reducer and virtual reality combination [7].

In general, Numbers of domestic scholars have studied the field of virtual roaming technology. This article takes the mechanical engineering training workshop of Shanghai Second Polytechnic University as an example to develop the scene roaming of the workshop practical training machinery.

2 System framework

First of all, the 3D model of all kinds of mechanical equipment is established. And the image processing makes the effect closer to the reality. The virtual workshop roaming system is developed under the Unity3D software. Finally, the virtual reality equipment zSpace is applied to realize the lathe simulation system with VR and AR functions. The specific development process is shown in Figure 1.

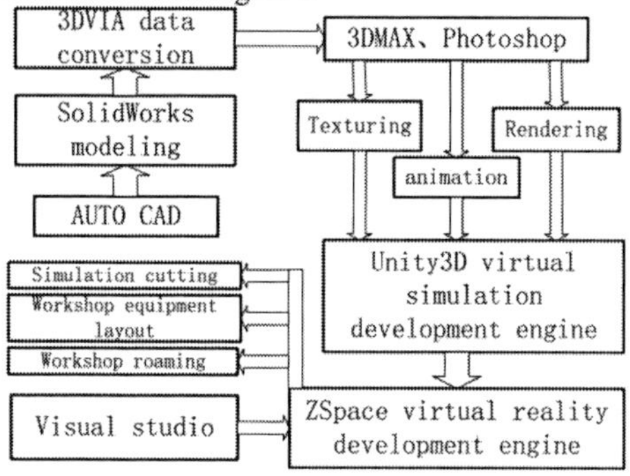

Figure 1. System framework.

3 Virtual factory model establishment

At the beginning of the system development, the first step is to do the survey and statistics of the actual workshops, especially the number and placement of all kinds of mechanical equipment in the workshop. table 1 is the number and names of all kinds of mechanical equipment.

Table 1. Factory equipment statistics

Device name	Ordinary lathe / table	Ordinary milling machine / table	Numerical control lathe/ table	Outer grinding machine / table
Number	12	5	10	2

3.1. Establishment of equipment model

Since Unity3D software itself does not have modelling engineering, SolidWorks software is used to establish a three-dimensional model of all kinds of equipment (Figure 2). All of the equipment size parameters are measured by actual measurement, which ensures the reliability of the model.

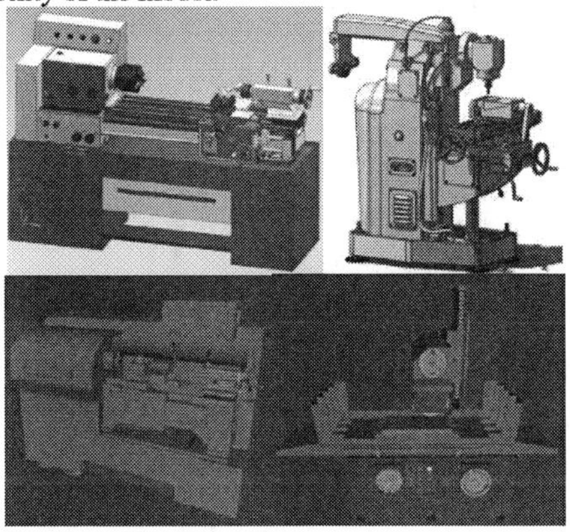

Figure 2. 3D model of mechanical equipment

3.2 Add texture and map

In order to make the virtual model closer to the actual model, the texture mapping operation of the model is carried out in detail. First, the surface pattern of the actual machine is photographed by the camera. The image is introduced into the Photoshop software for the correction processing as the material. The mapping operation is carried out in the SketchUp. Fig. 3 is a comparison of the map processing effect of the front and rear panels of ordinary lathe operation. It is obviously that the processed model is closer to the real effect [8].

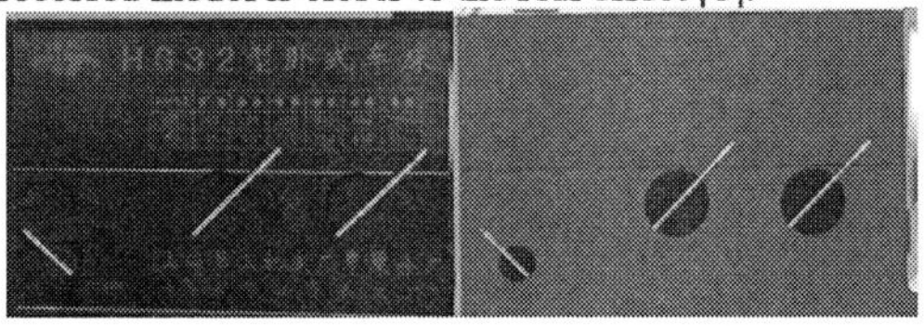

Figure 3. Comparison of the effect of map processing

3.3 Development of factory roaming function

Manual roaming is the most common virtual scene observation method. Through the upper and lower keys, the left and right movement of the model and the forward and reverse rotation are realized. The workshop roaming scene is shown in Figure 4. The typical program code is as follows:

```
if (Input.GetKey(KeyCode.A))
{
transform.Rotate(0, 20 * Time.deltaTime * speed, 0, Space.Self);
}
```

In order to clearly observe the model of mechanical equipment in the workshop, the roaming system designed the scaling and rotation functions of the model by calling the ray function and the Lerp function. The function call statement is as follows:

```
ray=Camera.main.ScreenPointToRay(Input.mousePosition);
transform.rotation=Quaternion.Lerp(transform.rotation, mRotation, Time.deltaTime * Damping);
```

Figure 4. Workshop roaming scene

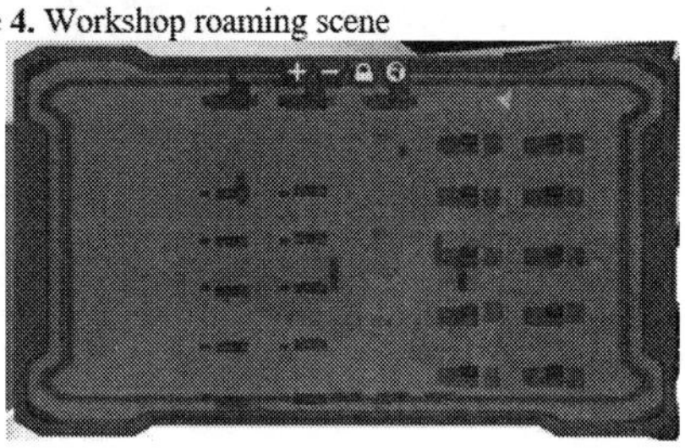

Figure 5. Small map technology

The workshop roaming usually requires the overall layout of the equipment. At this time the virtual camera cannot achieve the effect of the global observation. The system realizes the small map roaming technology in the workshop (Figure 5). The main functions are:

(1) It can display the orientation of the current scene in real time;

(2) It can reduce and put the size map to facilitate the rover to understand the current Location;

(3) Setting up map observation mode from two angles of facing and strabismus.

3.4 Development of plant equipment layout function

The rational design of the equipment layout is the key to the effective process of mechanical processing. The scientific and rational layout of equipment can improve the efficiency of the machining work, effectively save the production cost, ensure the quality and quantity of the mechanical processing, and also optimize the specific process of the production and processing.

Virtual factory design refers to the three-dimensional modeling of the workshop through the acquisition of physical workshops, production lines and logistics, and the building of a virtual manufacturing digital virtual workshop by using 3D design tools. Virtual factory layout has real-time and interactive layout function, and can be used for layout analysis in 3D scenario. Figure 6 is the free layout interface of the factory machine tool developed by the system. The mobile machine can automatically calculate the distance between the machine tool and the adjacent machine tools and display it in real time. Fig. 7 first select the scene for machine tool equipment. In the scene, not only the layout function is realized, but also the machine tool can be selected freely and the data is saved.

Figure 6. Machine tool layout scene

Figure 7. Machine tool selection scene

3.5 Development of virtual lathe simulation function

In the virtual simulation cutting scene, all the cutting technology and machine tool running data are collected from the actual cutting data, including the basic dimension parameters and the cutting parameters of the lathe. And the collected dynamic data are displayed in the processing scene by the way of mapping.

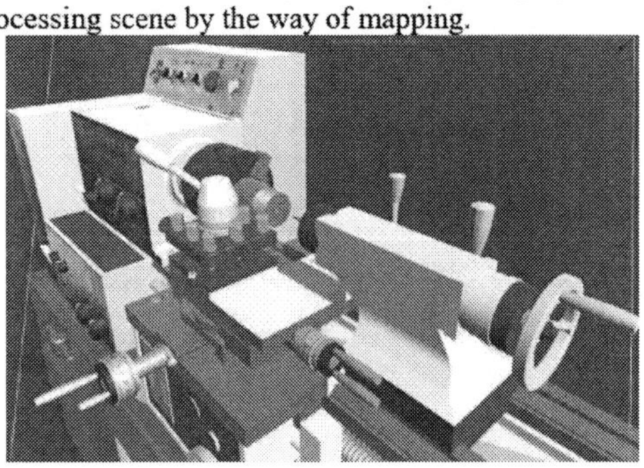

Figure 8. Simulation cutting scene

4 Development of virtual workshop roaming VR system

ZSpace system is a desktop virtual reality system that integrates real world work environment. The core technology is high fidelity stereo display system, low delay tracking system, stereo vision and direct interaction. Compared with the system developed by Unity3D, the virtual reality system developed by zSpace reflects the natural interactive and immersive of the new system.

4.1 Basic principles of virtual reality

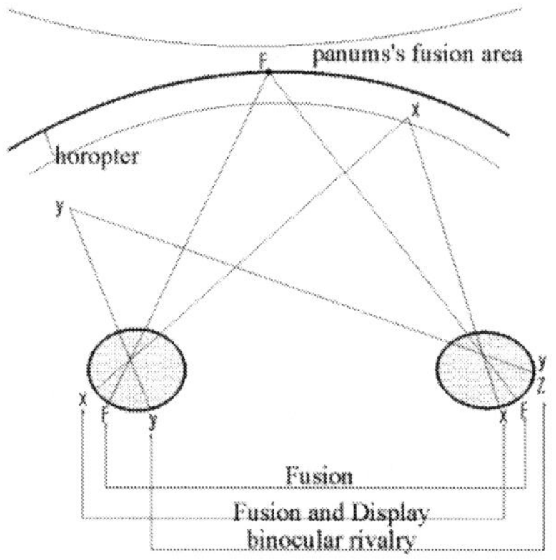

Figure 9. Stereoscopic imaging principle

Figure 3[9] describes the basic principle of zSpace stereoscopic display (Fig. 9). Assuming the fixed point F from the top down view, through the fixed point binocular (Panum fusion area), there are object X and object Y. When F is fixed, the F image stimulates the corresponding retinal points of two eyes and produces fusion. The object X is generated by cross parallax before binocular vision. Object X stimulates retinal differences in binocular fusion, as X falls within the fusion area of Panum. The object Y is in the front of the two eye view, with a cross parallax. The retinal difference point of the image from the object Y stimulates two eyes as double, because Y is outside the Panum fusion zone. Because Y has two retinal images and is in very different retinal regions, the Y image in the left eye may stimulate the retina region corresponding to one area in the right eye area. This image comes from the Z images of different objects in the field of vision, which leads to the competition of the binocular and forms the stereoscopic display effect.

4.2 Interactive design

The traditional system interaction design is realized by calling the UI function, which is, clicking the button by mouse. This is slightly intelligent compared with the traditional multimedia courseware. But it cannot make the learners feel the sense of being in the situation from the degree of interaction and immersion. This section attempts to call some functions in zSpace to implement the interaction between objects and objects. Figure 10 interacts with the model and model to roam the first view of the workshop (the system is a stereoscopic interface, and the picture is photographed).

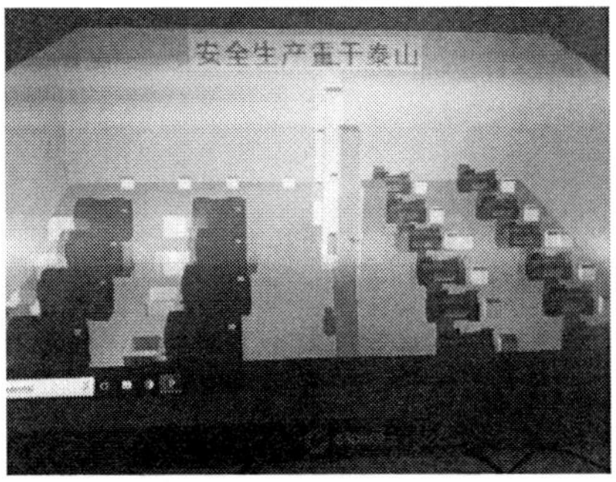

Figure 10. Total system interface

4.3 Application of Augmented Reality Technology

Augmented reality is a technology that combines real environment with virtual reality. Augmented reality system is usually designed for human vision, with graphic images as the main output media. Achieving more natural interaction between users and virtual objects in real scenes is one of the goals of augmented reality technology. Therefore, the display function and interaction function in augmented reality is the key to reflect the application value of the system.

This system has the function of virtual reality, and it also has the function of augmented reality. The zView device can share the stereoscopic model seen in the VR. It uses the high definition camera as the hardware, screen overlapping technology as software, to overlap the zSpace picture. The purpose of the technology is to screen the virtual world in the real world and interact with it. Fig.11 is the augmented reality effect of the main structure of the lathe and gear in the virtual workshop.

Figure 11. Augmented reality

4.4 Conclusion and Prospect

With the development of virtual reality technology, this technology has been applied in engineering and technology fields often and often. This system aims at developing virtual factories in the field of intelligent manufacturing. First, the system development framework is formulated, the function of the system is determined; the functions of virtual factory roaming and virtual equipment layout are developed. The system can quickly understand the layout of the plant equipment, as well as helping the factory managers to optimize the layout workshops reasonably. The production equipment has played an important role in saving time and improving efficiency. At the same time, in order to reflect the visual operation interface, the system becomes more interactive and immersive with the aid of virtual reality operation engine. The engine also provides some reference for the design of intelligent factory.

The follow-up study combines the virtual reality glove with the machine tool manufacturing process to develop a more advanced interpersonal interaction, which makes the virtual reality technology better serve the intelligent manufacturing process.

Acknowledgements

This work was supported by the Key Discipline Project of Shanghai Polytechnic University (No. XXKZD1605) and Graduate project fund of Shanghai Polytechnic University (No. A01GY17F022).

References

1. Shi Min, Wang Junzheng, Wei Jiahui. Realistic 3D virtual scene construction and roaming method [J]. Journal of System Simulation,2014,(9):1969-1974,1979.
2. Zou Hong, Yang Hong, Cui Xiaowei, et al. Virtual simulation scene and interactive roaming system [J]. Oil-Gasfield Surface Engineering, 2014, (1): 22-23.
3. Hu Fuwen, Chen Chengkun, Liu Yancheng, Li Li. Research on virtual planning system for product assembly process [J]. Mechanical design and manufacture, 2016, (12):180-183.
4. Yu Qiang, Li Xiangsong. Application Research of digital plant layout simulation technology [J]. Machinery & Electronics, 2015, (11): 21-24.
5. Wen Huaixing, Huang Zhengxiang. Research on Virtual Assembly System Based on Virtools parallel machine tool [J]. Manufacturing Technology & Machine Tools, 2017, (9): 91-94.
6. Liang Hongbao, Cheng Qi chao. Optimization of 3D model of virtual factory based on Virtools [J]. Computer and Digital Engineering, 2008,36 (11): 166.
7. Zhu Wenhua, Cai Bao, Shi Kun Ju, Liu Sai. Research on the disassembly and assembly of reducer combined with virtual and real [J]. Laboratory Research & Exploration, 2017,36 (11): 92-108.
8. Hoai Nam Le Tran, Kabilen Sornum. A Systematic Approach for Rapid 3D Reconstruction from Photo Sets. 2010 11th Int. Conf. Control, Automation, Robotics and Vision Singapore. 2010(12): 1167-1174.
9. Patterson, R. Human factors of stereo displays: An update. Journal of the Society for Information Display, 2009(17): 987-996.

Methodology for Enhancement of Energy Efficiency in Hydraulic Equipment

Sergio Luiz Pereira, Guilherme F. Fa. Sortino, Eduardo Mário Dias and Maria Lídia Rebello Pinho Dias

Department of Electrical Energy and Automation, University of Sao Paulo – Polytechnic School - Sao Paulo, Brazil

Abstract. The cost of energy has increased greatly in the last decades, even with technological impact attenuators, which brings a concern with the increase of production effectiveness and better use of energy by machines and advanced equipment in industrial plants. This article proposes a methodology that allows a more rational use of energy in the production environment, applying Automation Technology, which represents a more affordable approach in relation to the high capital demanding complete machine replacement by the latest highest efficiency version available. The methodology thus contribute to reducing capital consumption, reducing costs, impacting both in the productive and financial performance of the organizations, with the overall result meaning less damage to the environment, providing sustainable gains.

1 Introduction

The changes in the world economy in the last decades have had substantial impact on the industrial sector, as the Third Industrial Revolution, led to the modernization and automation of production processes, accelerating the manufacturing systems and allowing increase the quantity produced and the product diversification (VASCONCELLOS, 2013) [1]. Many industrial equipment, at least with regard to energy use, were kept in the same technology standard since they were created, some of which even have added new energy demands by incorporating technological innovations and new automation devices. After the 1990s, with the establishment and consolidation of the age of the New Economy and, more recently, in the age of the Information Society, it has been possible to see the genesis of an original approach, called "Ecoeconomics", whose foundations are in Pereira and Simão Filho (2014) [2], based on the proposals of Georgescu-Roegen, Herman Daly et al, originally made in the 1970s, where they show that the society and the economy are undergoing evolutionary steps, often of the organizational and technological character, which promote changes affecting the form of economic organization and also the physical equipment of industrial processes. The changes, including in the decision-making processes, have incorporated, ultimately, the notion that we seek constantly "improving the efficiency of all processes" with the economic system favoring essentially the human being (PEREIRA; SIMÃO FILHO, 2014, p. 15) [2]. Al Gore (2009) [3], for example, mentions that after the human travel into space in the 1960s, when the planet Earth could be seen in its entirety, it became clear the notion of planetary limited resources, to the point that the UN (United Nations) proposed since then meetings to discuss the concern with the limits of resources and respect for the environment. Thus, it is projected on the business environment, especially in the evolving Century XXI, a growing concern for ecosystems, with the limitation of production factors and energy efficiency, so that the rational use of energy passes to be seriously considered in corporate decision-making processes, first for the equipment and light and flexible machines, and then to the large industrial equipment (SCHULER, 2014) [4]. This paper presents the MAAMEES (original acronym in Brazilian Portuguese) Methodology for the Application of Automation in Large Industrial Equipment for Enhancement of Energy Efficiency and Sustainability Gains, proposing a developed solution to help the industry become more energy effective, with gains not just for the economy but above all for the whole society and the environment.

2 Methodology for energy effectiveness

The processes of transformation of an organization are benefited by the adoption and application of appropriate methodologies (SILVA, 2013) [5]. For the purpose of this article there is the Methodology for the Application of Automation in Large Industrial Equipment for Enhancement of Energy Efficiency and Sustainability Gains, known as MAAMEES. Sustainability Gains, know as MAAMEES. Sustainability Gains, known as MAAMEES. Sustainability Gains, known as MAAMEES. The principles of the use of MAAMEES are based on the proposal of use of AT with the objective of eliminating a significant percentage of the energy expended without the effective transformation into useful work that the traditional operation of large hydraulic machines perform. This means the search for the reduction of the energy expenditure in front of the total work done, meaning in practice an increase of energy efficiency. MAAMEES consists of some fundamental steps that include: i) "ex-ante" analysis; (ii) implementation; and (iii) "ex-post" analysis. The flowchart, shown in figure 1, is recommended for its application and allows understanding the process.

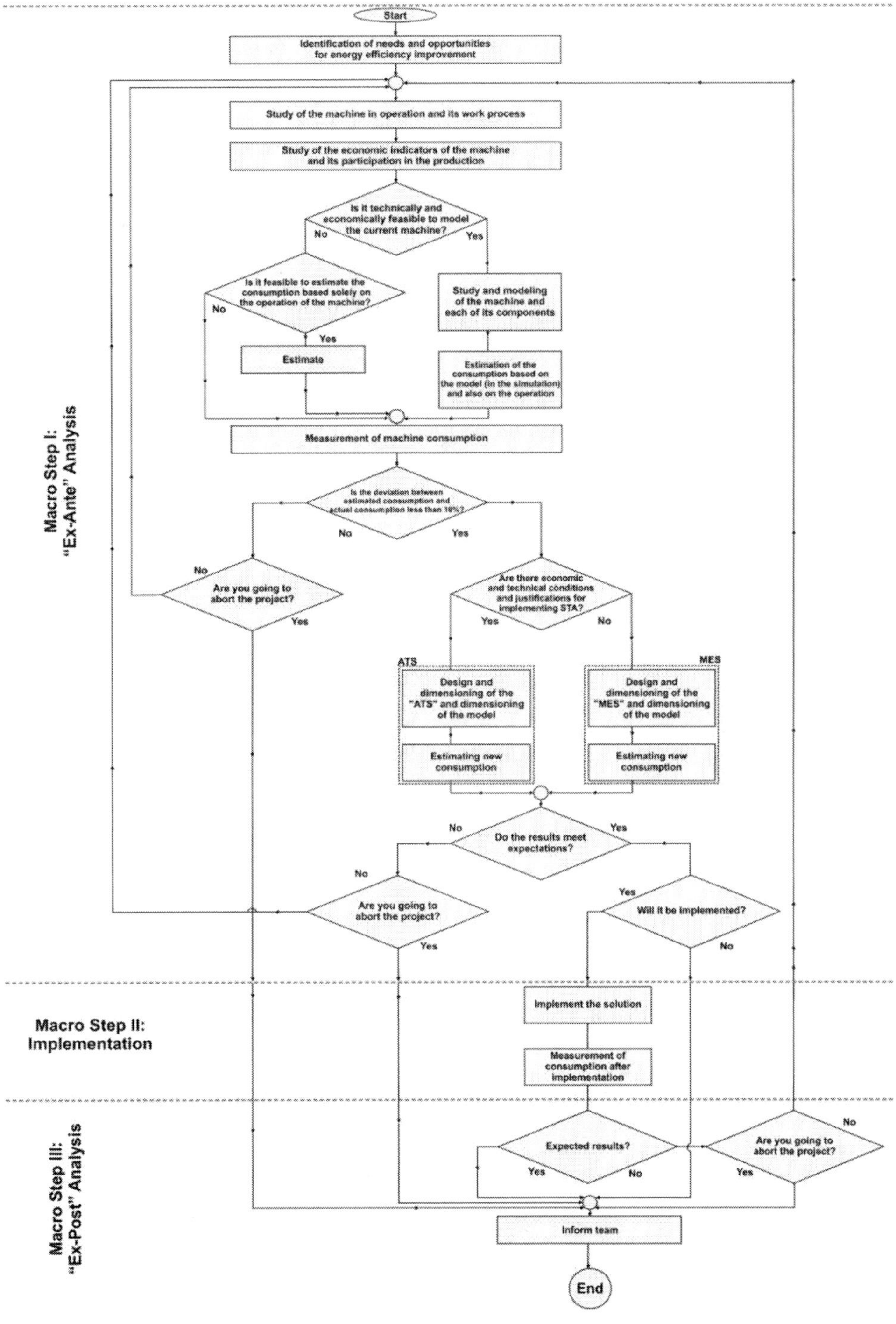

Figure 1. MAAMEES Flowchart.

In the "ex-ante" analysis, a preliminary and detailed diagnosis should be made of the unit where the process for energy efficiency will be applied. At the same time, it is important to analyze the current situation of the company regarding the concern with the energy economy and with environmental and sustainability issues. Also, an analysis of the current contribution of the industrial unit in terms of energy economy and its degree of commitment to the environmental and sustainability issues of the company as a whole, as well as an analysis of the industrial processes and machinery and equipment used in the industrial unit regarding energy efficiency. It is important to carefully identify the equipment or machine object of the application of the methodology for energy efficiency gains. It is desirable, prior to the application of the methodology, to study in advance the environmental, technological, human and economic financial viability of the unit that will suffer the direct impact of the Methodology and resulting interventions. A diagnosis of the AT situation in the company and in the industrial sector should also be made based on corporate and sector reports that express the commitment of the company and its competitors to the use of automation for industrial production. This analysis aims to identify if the company has already assimilated and if the companies of the sector have already assimilated the latest automation technologies. Also, an analysis should be made of the situation of the company and its sectoral competitors in terms of commitments to energy efficiency and sustainability. It is advisable to identify as thoroughly as possible what actually exists and what has already been planned with a view to energy efficiency in the industrial sector of the company. It should then analyze what the existing automation produced from its adoption, thus, identifying the situation to know if the existing level of automation already contributes in some way to energy efficiency. It is also necessary to identify whether the current automation technology has impacts on the environment, human resources and other factors of interest, such as stress and noise. In this step, a survey is also made of indicators that measure the main parameters of interest, particularly the energy consumption of the process studied and the effects on production capacity. Next, it is necessary to analyze in advance the possible efficiency gains provided by the different solution alternatives that exist and that are economically feasible within the industrial, product and market conditions to which they belong. A virtual analysis is recommended through technologies of early simulation of the potential results after the changes that lead to the gain of energy efficiency are implemented. A study should be made of the investments required for the alternative that is most appropriate for the existing context, as well as anticipated studies on possible impacts and expected results, and the analysis of environmental, technological, human and economic-financial viability. If possible, the theoretical model should be used to measure a priori the energy efficiency gain for the modifications to be implemented, making the recommendations for administration as the preliminary studies indicate that the path is feasible. Once feasibility has been obtained and the necessary approvals have been obtained, the whole process of implementing the modifications chosen must be monitored, so as to ensure that they are carried out as planned. Given that the proposed changes were decided and the implementations were carried out, measurements and indicators should be collected, which will allow the analysis of the gains actually obtained in terms of energy efficiency, enabling them to be bought with the ex-ante situation. Similarly, it should be verified whether the theoretical gains in energy efficiency have been confirmed by comparing the actual results before and after the implementation of the methodology, identifying and confirming the energetic efficiency gains predicted and measured through actual results. After confirmation of the energy efficiency gains, one must also identify the gains in terms of human consumption, input savings, CO2 emissions, among other economic, environmental and sustainability factors.

When analyzing the initial macro phase of MAAMEES, the "Ex-Ante" Analysis, empirical situations show that there are two possibilities for the improvement of the large industrial hydraulics machines, which are called ATS (Advanced Technological Solution) and SME (Solution of Maximum Effectiveness). It is the economic analysis of the existing conditions that determines whether to adopt the ATS or SME. The ATS involves the utilization of a structural part of the equipment, performing its modernization through investments in the replacement of the pre-existing conventional hydraulic system by a modern servo motor coupled to the hydraulic pump of high precision and efficiency, which are integrated into dedicated management of processes and operations. In other words, if the hydraulic machine in question is used to produce volume that represents high added value in the production process, it is justified to use the ATS solution, which will require additional costs in the acquisition of modernization equipment and whose main suppliers currently in the world are the German companies Voith and Siemens. The ATS application results in the use of less physical volume by its components in the machine than the conventional hydraulic solution being replaced. This feature facilitates the modernization processes of equipment. The sensors used in the modernization process by ATS also favor appropriate diagnostics for the application of preventive and predictive maintenance. The SME solution, on the other hand, is indicated when the aggregate value and volume of production is not high and therefore it is not justified to make additional investments in equipment modernization assets. In this case, an update can be made to the safety device control software of the hydraulic machine, making the time the machine stays on being reduced and leading to energy efficiency gains. When adopted, the SME may have two basic architectures: when the machine already has an on-off control system managed by a programmable logic controller, no spare equipment will be needed. When the machine is old, a programmable controller has to be integrated to the drive system and position sensors must be attached. The SME solution has as principle the automatic shutdown of the hydraulic machine when it is not in production, seeking to take advantage of existing solutions, such as functional safety at the operation site, and it is relatively cheaper because it requires only the analysis of the production process and the machine itself, with the insertion of new logical sequences in the existing AT software. This solution takes advantage of the equipment, adding additional routines of automation through adjustment and insertion of command lines in the software that are responsible for adjusting the productive demands to the potential of reduction of consumption identified in the study of the equipment. It also improves the production process in order to ensure the operation of the hydraulic system in the minimum feasible period for the execution of the scheduled batches. This logic allows the machine equipped with the SME to remain disconnected in most of its interval periods or idleness between the processing activities of the batches produced, with a minimum of interference from the operators. When applying the MAAMEES, whether in the ATS or SME version, double direction gains are observed, since the company will benefit from the positive influence of the real Eco-Economic environment, which is also inserted in the achievement of more efficient energy indicators, as well as will be the contribution of the company to the economy as a whole, where companies from the same value chain can also benefit from the applied methodology.

3 The empirical test of MAAMEES

From the methodologies presented above, MAAMEES stands out for its effectiveness in achieving energy gains, first because it is relatively simple and second because it presents objective results as soon as its implementation occurs. MAAMEES was submitted to two empirical tests, the first one, using the SME (Solution of Maximum Effectiveness) version, was carried out in a hydraulic machine in the MBB Mercedes Benz do Brazil, and the second one, using the ATS (Advanced Technological Solution) version, applied to a Voith's client machine in Germany.

4 The SME approach

The plant has several hydraulic machines of different generations, requiring engineering decisions aimed at improving energy efficiency. The application of MAAMEES followed the steps defined in the flowchart shown in figure 1, containing the Macro Phases I, II and III, respectively "'Ex-ante' Analysis", "Implementation" and "'Ex-post' Analysis", the last two carried out by simulation. A preliminary diagnosis was made at the MBB's operational unit, where tens of presses with capacities of up to 1,300 tons, several of them hydraulic, were in operation, among them the Ciola 60t Press (manufactured in 1976), model 1C-60BA, with a 250 x 250 mm table, 150 mm stroke, 300 mm opening, mono speed and two-hand trip, and modernized in 2014 to meet Brazilian safety standards. The machine was studied during a typical production week when an average of 336 pieces per day was produced, that is 168 pieces per shift or even 21 pieces per hour on average. The production made by this machine is of low value added and it would be necessary to invest approximately €20.000 for its modernization within the ATS concept of MAAMEES. The company's executives chose the simplest solution, updating the machine's operating software within the MAAMEES SME concept. During the 120 total measurement hours, the machine operated for about 80 hours, in two shifts, with the average energy consumption per hour of 2.03 kWh, producing 1,680 parts or 336 parts per day, obtaining 10.34 parts per kWh. In terms of potential of production, the machine can produce 360 parts per hour, or 177.3 parts per kWh, which shows that the energy efficiency measured is approximately 94% lower than it could be if the machine were fully used or if it was disconnected during the periods of non-production. The Machine is therefore used only 6% of its operating time. With the implementation of SME the machine is switched off after 90 seconds of idle operation. The process of arriving pieces is given by crates containing in average thirty pieces, with great randomness of arrival. Due to the variability of the parts arrival process, it is not possible to estimate the energy consumption accurately, thus estimating the possible energy savings based on two extremely demanding machine behaviors: scenarios A and B shown on table 1. **Scenario A:** Constant Request Behavior Based on Average Production. The behavior of the machine in this scenario assumes a constant ratio of the arrival rate of 21 pieces per hour, or approximately 1 piece per 171 seconds. By adopting for the purpose of this study that this behavior of the productive plant between the arrival of one piece and another is correct, there will be an interval of 171 seconds between one piece and the next one. As the proposed control algorithm commands the shutdown every 90 seconds of waiting, in this hypothetical operational situation the machine would turn off something around 40% of its normal production time. In this estimation we have already considered the stabilization times during the drive. It is important to point out, however, that this scenario is hardly feasible in the practical reality of the industry.

Table 1. Possible energy saving scenarios with SME deployment

SME Implementation	Reduction of Energy Consumption	Economy in KWh (Average per Hour)
Scenery A	40%	0,81
Scenery B	94%	1,91

Scenario B: Behavior of Requests with the Arrival of the Whole Lot with the Full Volume of the Daily Schedule. In this scenario it is considered that 336 pieces arrive at almost the same time. In this way, a queue is formed where the pieces are processed one after the other. As the machine capacity is 360 parts per hour, that is, 10 seconds per piece, it will take 3,360 seconds or 56 minutes of operation for the production of all 336 pieces. In this way the machine would only be turned on for about 1 hour from the 16 hours of the two shifts. That is, only 6.25% of the production time would be connected. If the total consumption of the machine every two shifts is

32.5 kWh, the estimated hourly consumption in this case would be 2.03 kWh. Thus, considering scenario A and scenario B, the range of consumption reduction would be between 40% and 93.75%. Table 2 illustrates the energy gain data for the two scenarios with SME deployment. It should be noted that the expected results for the two cases are very satisfactory, so that the executives responsible for decision making process in the company have opted definitively for the implementation of SME. The company considered the results very important, so that this engineering project was adopted by the organization's energy efficiency schedule. Initially, the machine demanded almost 7.9 MWh per year in its operation, at a cost of more than R$3,173.00, but with the application of MAAMEES and particularly its SME approach, it was possible to reduce this energy consumption to only 0.79 MWh per year, providing a monetary savings in the order of more than R$2,856.00 for one machine only. From an investment of just over R$4,600.00 the company obtained in terms of reduction of energy consumption a cost savings of more than R$2,856.00 per year for a machine with a useful life of more than 10 years.

Table 2. Actual and simulated results on the studied machine

Results for the Studied Machine	Energy Consumption in 365 days (MWh)	Annual Energy Cost (R$)	Savings (R$)
Situation *ex-ante*	7,86	3.173,87	-
Situation *ex-post*	0,786	317,39	2.856,48

With this data it is possible to carry out the economic viability analysis of the implementation of SME in one or more similar machines. For this purpose, the methods of valuation Payback Period, Net Present Value and Internal Rate of Return will be used. In the calculus was considered the interest rate published by the Central Bank of Brazil for February/2017, which stood at 13% per year. By making a linear extension of the result found in the simulation for one machine, it was possible to already see from five machines that the company has its costs with the SME diluted and it obtain a positive and continuously increasing net result. Applying SME to five similar machines, the company obtains a gain of R$7.7 thousand per year, which reaches more than R$243 thousand when extending the application of SME to 100 similar hydraulic machines.

Table 3. Income statement for simulated scenarios

Scenery			Annual Revenue	Cost	Net Result
1	1	conventional hydraulic machine	R$ 2.856,48	R$ 4.468,25	-R$ 1.831,77
2	5	similar machines	R$ 14.282,41	R$ 6.563,55	R$ 7.718,86
3	10	similar machines	R$ 28.564,81	R$ 8.438,85	R$ 20.125,96
4	50	similar machines	R$ 142,824,06	R$ 23.441,25	R$ 119.382,81
5	100	similar machines	R$ 285.648,12	R$ 42.194,25	R$ 243.453,87

For the "Payback Period" method, both inflation and interest rates are of no importance, since the analysis is only nominal, and it shows how long the company can recover the investment made in the application of the SME. In table 4 it is possible to see that the resources used in only one machine would take almost 20 months to be recovered, but for 5 machines the time of recovery of the investment falls to less than 6 months, showing full feasibility in the application of SME from the point of view of the Payback method. It is important to note that in the scenario where SME is applied for a park of 100 similar hydraulic machines, the Payback Period is reduced dramatically to less than two months.

In table 5 it is possible to see the nominal cash flow for a period of five years, applied to each of the five scenarios defined. The application of the SME to only one machine already presents a positive Net Present Value (in an economy with a 13% annual interest rate) and an Internal Rate of Return of 54% a year, which shows the full feasibility for its implementation, under the conditions considered in the simulation.

Table 4. Viability analysis by payback period

Scenery	*Payback* Period	
	Month	Year
1	19,7	1,64
2	5,51	0,46
3	3,55	0,30
4	1,97	0,16
5	1,77	0,15

Table 5. Simulated cash flows, NPV and IRR results

Scenery	N.P.V.	I.R.R.	Machines	0	1	2	3	4	5
1	R$ 4.742,17	54%	1	-R$ 4.688,25	R$ 2.856,48	R$ 2.856,48	R$ 2.856,48	R$ 2.856,48	R$ 2.856,48
2	R$ 64.848,48	217%	5	-R$ 6.563,55	R$ 14.282,41	R$ 14.282,41	R$ 14.282,41	R$ 14.282,41	R$ 14.282,41
3	R$ 134.385,21	338%	10	-R$ 8.438,85	R$ 28.564,81	R$ 28.564,81	R$ 28.564,81	R$ 28.564,81	R$ 28.564,81
4	R$ 690.679,05	609%	50	- R$ 23.441,25	**R$ 142.824,06**	**R$ 142.824,06**	**R$ 142.824,06**	**R$ 142.824,06**	**R$ 142.824,06**
5	R$ 1.386.046,35	677%	100	- R$ 42.194,25	**R$ 285.648,12**	**R$ 285.648,12**	**R$ 285.648,12**	**R$ 285.648,12**	**R$ 285.648,12**

When the application of the SME for each of the five scenarios is extended, the situation improves significantly, with the NPV reaching almost R$1.4 million for the case of 100 machines and IRR of 677% per year. The results obtained prove that the application of the SME is feasible not only for a machine but for different scenarios where it would be applied to a huge number of similar hydraulic equipment in the industrial plant.

5 The ATS approach

The application of MAAMEES in its ATS (Advanced Technological Solution) version was done by simulation in an industrial plant of a Voith's client in Germany, where a 4,000 kN press was chosen. The hydraulic press is allocated to produce parts of three different types of products: A-type parts, B-type parts and C-type parts. The constructive details and operational characteristics of the three types of parts were not considered relevant for the empirical test. However, it is known that the piece of type A demands 9.33 seconds to be produced, the piece of type B demands 4.5 seconds and the part of type C demands 9.0 seconds, according to table 6.

Table 6. Nominal capacity in time by type of parts

Period / Type of parts	Hour	8 hs shift	Day/3 shifts	Week/5 work days	Year/50 work weeks
A	385	3.080	9.240	46.200	2.310.000
B	800	6.400	19.200	96.000	4.800.000
C	400	3.200	9.600	48.000	2.400.000

Source: Adapted from Voith (2017) [6]

The productions for the year 2016 of the three pieces, type A, type B and type C, were respectively: 577,500; 1,200,000 and 600,000 units of each type. It is known that the average energy demand of this hydraulic machine is 11 kW continuously. During the data collection period the machine was running three shifts a day, i.e. the machine remained on 24 hours a day and 120 hours a week. It was observed that, regardless of the type of part produced, the weekly consumption was 1,320 kWh, meaning that the press consumed 264 kWh per day, 88 kWh per shift or 11 kWh per hour. These measurements were carried out throughout 2016 when the machine energy annual consumption was 66,000 kWh. Production and energy consumption data indicate three productivity indices: 8.8 pieces per kWh for Type A parts, 18.2 pieces per kWh for Type B parts and 9.1 pieces per kWh for type C. Although the press was operating at 100% of its operational capacity, the energy indicators would be: 35 pieces per kWh, 72.7 pieces per kWh and 36.4 pieces per kWh, respectively. This shows that the energy efficiency of this press for the three types of parts is around 25%. In the case of the Voith customer machine in Germany the volume of production and the value added of the production is undoubtedly bigger than in the case of the press studied at MBB. So the control routine that shuts off the machine when the time between parts is bigger than 90 seconds does not meet the expectations of reduction of energy consumption. Due to the high volume of parts produced by the press and its added value, the project management team opted to apply the MAAMEES in its ATS version. With the application of ATS, the press, when connected, works only in its electronic part, remaining with the step motor inactive until it is activated by the operator when executing an operation of pressing. The motor shuts off as soon as the operation is completed, with only the electronic part of the machine switched on until a new part is positioned and the process is started again by the operator. With the implementation of ATS, the press request by the planning and production control sector remains the same. However, the electrical power demanded by the stepper motor is only observed at the actual moment of the mechanical work of the press. The total cycle analysis allows the conclusion that the pressing, stamping and pressure retention times correspond on average to 15% of the time of a pressing cycle considering the usual waiting times. These times vary according to the requests of each piece. Thus there is an energy saving space if the stepper motor only operates at the times of effective work. As the times vary greatly and as the machine with ATS stays on, there is no possibility to estimate the very precise consumption reduction. However, it is possible to assume that the estimated reduction of energy consumption should be between 60% and 80%. As the consumption reduction data were very favorable throughout the project, the management group decided that the ATS alternative should be implemented. The ATS was put into operation and it was checked whether the mechanical performance for the production of parts remained the same. A strict quality control and compliance with the new parts produced were also carried out. Once the ATS alternative was approved from an operational point of view, the electric consumption meter was re-installed to carry out new measurements and the comparison between estimated consumption and actual consumption. These measurements were carried out during a month in the first half of 2017 and the data of the new consumption are shown in table 7.

Table 7. Energy efficiency results with ATS application

Parts type	Energy Consumption	Reduction
A	1.350 kWh	75%
B	800 kWh	86%
C	1.367 kWh	76%

Source: Adapted from Voith (2017) [6]

It should be noted that the consumption reduction was within the expected range, with 75% for part A, 86% for part B and 76% for part C. The monthly volume of production during the measurements was practically the same and energy efficiency of the press with the ATS has achieved significant gains, going to, in the case of part A to 35 parts per kWh, for part B, 73 pieces per kWh and for part C was 36 pieces per kWh. The original conditions considered for calculating the energy efficiency improvement of this press while operating with its conventional hydraulics is approximately 11 kW continuously. In addition, the following working parameters are considered: three shifts, five working days per week and 50 weeks per year. ATS's application resulted in the average energy cost per unit produced in the operation for cycle 1 of € 0.54, for cycle 2 of € 0.32 and for cycle 3 of € 0.52.

The organization's expectations that MAAMEES could contribute to the reduction of the average operating cost of energy, while preserving the production volumes unchanged, have been confirmed since the hydraulic machine in its conventional configuration results in an annual energy consumption of 66,000 kWh, which represents an annual cost of €13,200.00.

Table 8. Results of energy consumption, cost reduction and payback with ATS application

Parts type	Annual Energy Consumption	Annual Saving	*Payback* period
A	16.200 kWh	€ 9.960	1,75 years
B	9.600 kWh	€ 11.280	1,5 years
C	15.600 kWh	€ 10.080	1,7 years

Source: Voith (2017) [6]

As can be seen in table 8, the application of MAAMEES in the ATS version led to a reduction of costs for all parts studied, with an annual energy saving of €9,960, € 11,280 and € 10,080 respectively for type A, B and C. The original calculations, on a per-piece basis, showed that these cost reductions give the company a Payback Period, facing investments made versus realized gains, of 1.75 year for the manufacture of A-type parts, of 1.5 year to manufacturing of B-type parts, and 1.7 year for the manufacture of C-type parts. These Payback results are significant and prove, in principle, the viability of the ATS application. However, the team of engineers who conducted the project decided to apply a scenario analysis similar to that used in the experience with SME at MBB, allowing an economic view on the application of the methodology when it extends to industrial parks that have many hydraulic presses similar in operation. The cost of implementing of the ATS solution for a single machine is €17,172.00 and when a simple results analysis is performed for a single machine, this investment is compared with an average annual revenue (calculated by the weighted average of the annual savings shown in table 10) of €10,331.00, which can give the impression that the company would suffer losses, as the economic result would be negative (€ 17,172.00 is invested and a gain of € 10,331.00 is obtained, with a loss of € 6,841.00) However, the economic analysis must take into account the time, requiring the use of cash flows, since the machine will receive investments and will continue to work for a long time - and not just for a single year. Conservatively, for example, both the SME and the ATS have been studied for a 5-year planning horizon, although in many cases, due to the effectiveness of the industry maintenance programs, the hydraulic presses can operate several years and even decades. Another relevant aspect to be specifically considered in the case of the ATS application is the fact that new equipment from an external manufacturer is purchased to modernize the existing hydraulic machines. In the case of SME, only to counter, investments occur in existing software, not involving the acquisition of additional equipment from external manufacturers. This means that by applying to ATS, it is possible to negotiate discounts on the price paid for the equipment when the company wants to modernize 5, 10 or 50 hydraulic machines of a certain industrial park. The ATS application engineers consulted the equipment manufacturer and obtained the results summarized in Table 9, with staggered discounts for the application of ATS in increasing numbers of machines.

Table 9. Scenarios, revenues and investments for ATS application

Scenery			Annual Revenue	Investment
1	1	Conventional hydraulic machine	€ 10,311.30	€ 17,162.00
2	5	Similar machines	€ 51,556.52	€ 81,519.50
3	10	Similar machines	€ 103,113.04	€ 157,890.40
4	50	Similar machines	€ 515,565.22	€ 772,290.00
5	100	Similar machines	€ 1,031,130.43	€ 1,458,770.00

It should be noted that the value of the investments does not increase linearly, since the manufacturer may grant discounts for the acquisition of different numbers of equipment for the application of the ATS. The annual revenue, however, calculated by the weighted average presented in Table 10, increases linearly, since it is the direct technological result of the application of the MAAMEES in the equipment studied. The cash flows for the five scenarios studied can be seen in Table 11, where the investments occur at the initial moment (year zero) and the results are accounted for at the end of each year. With these values it is possible to calculate the economic indicators that show if the application of ATS in the different scenarios will be viable.

Table 10. Cash Flows for ATS application in different scenarios

		Year and value in €					
Scenery	Number of machines	0	1	2	3	4	5
1	1	17,162.00	10,311.30	10,311.30	10,311.30	10,311.30	10,311.30
2	5	81,519.50	51,556.52	51,556.52	51,556.52	51,556.52	51,556.52
3	10	157,890.40	103,113.04	103,113.04	103,113.04	103,113.04	103,113.04
4	50	772,290.00	515,565.22	515,565.22	515,565.22	515,565.22	515,565.22
5	100	1,458,770.00	1,031,130.43	1,031,130.43	1,031,130.43	1,031,130.43	1,031,130.43

One of the first indicators calculated by the team of engineers accompanying the ATS project in Germany was the Payback Period, now considering the various scenarios, as shown in Table 11.

Table 11. Payback for the different ATS application scenarios

Scenery	Number of machines	*Payback* (years)
1	1	1,66
2	5	1,58
3	10	1,53
4	50	1,50
5	100	1,41

The observed Payback indicators closely approximate the values presented previously, in the aggregate analysis, without considering scenarios, which also showed feasibility with the return of the investment occurring approximately 1.5 years after the beginning of the application of the ATS methodology. The engineering team, however, also calculated the indicators of Net Present Value and Internal Rate of Return, as can be seen in Table 15, completing the feasibility analysis. Specialized websites in Europe clarify that interest rates in Europe are still very close to zero, but in everyday business practice it is necessary to use an opportunity cost between "investing" and "lending" money at the current maximum rate, hence why the companies use 5.30% per annum as interest rate to calculate the feasibility indicators (see www.tradingeconomics.com/euro-area/interestrate). The ATS project engineers used this rate to determine the results presented in Table 12, where it can be seen that, even for just one machine, the ATS application is already feasible, generating a net return of more than € 25,000, within a 5 year analysis horizon. The same is true for a larger number of machines in the different scenarios, for example NPV of € 270 thousand generated for the application of ATS to 10 machines and NPV of € 2.82 million for the case of an industrial park of 100 machines.

According to the NPV criterion, as it was already the case with Payback, the ATS's economic and financial viability is already fully demonstrated. With the Internal Rate of Return calculation, the engineers show that the implicit return on operations (53% to 65% per annum) is well above the rate of opportunity cost in the Euro Zone (5.30% per year), completing the viability.

Table 12. NPV and IRR for the different ATS application scenarios

Scenery	Number of machines	Net Present Value (€)	Interest Return Rate (% per year)
1	1	€ 25,748.10	53%
2	5	€ 132,815.04	57%
3	10	€ 270,519.55	59%
4	50	€ 1,368,895.93	60%
5	100	€ 2,819,282.84	65%

References

1. VASCONCELLOS, M.A. S. Economia: micro e macro. São Paulo: Atlas, 2013.
2. PEREIRA, S. L.; SIMÃO FILHO, A. A Empresa Ética em Ambiente Econeconômico: a contribuição da empresa e da tecnologia da automação para um desenvolvimento sustentável inclusivo. São Paulo, Quatier Latin, 2014.
3. GORE, A. Our Choice: how we can solve the climate crisis. New York: Bloomsbury, 2009.
4. SCHULER, A. G. Energy optimization for press systems. Berlin: Powersave, 2014.
5. SILVA, E. B. Metodologia para Planejamento da Convergência da Tecnologia da Informação (TI) & Tecnologia da Automação (TA) em Processos Industriais. Tese de doutorado. São Paulo: Escola Politécnica da Universidade de São Paulo, 2013.
6. VOITH TURBO H + L HYDRAULIC GMBH & CO. KG. PSH Press Drive Servo Hybrid: servo pump drive system for hydraulic presses. Rutesheim, 2017.

Research on Operation Analysis and Evaluation Method of Ergonomics for Virtual Assembly

Guobin Gou[1], Xu Zhang[1] and Yuehui Yan[2]

[1]*School of Mechanical Engineering Beijing Institute of Technology. Beijing 100081, China*

[2]*China Academy of Launch Vehicle Technology. Beijing 100076, China*

Abstract. With the rapid development of digital technology and virtual reality technology, virtual reality technology has been paid more and more attention by more and more people. To investigate the application of virtual assembly simulation and improve ergonomics, the study on the interactive assembly operation has been performed. A Comprehensive Evaluation Method for ergonomics based on simulation of virtual assembly operations is proposed, aiming at avoid inadequate working postures, effectively improve assembly quality and shorten the production cycle. In this paper, the virtual assembly technology is introduced to the assembly process of an industrial manipulator. Details of the virtual assembly simulation process of an industrial manipulator design and its implementation are presented. This proposed method provides great support to the design of assembly process and brings about good change in the ergonomics of machine and system visualization. The analysis of this instance plays a guiding role in the optimization of human factors engineering.

Keywords: virtual assembly, assembly operation, ergonomics, evaluation.

1 Introduction

To succeed and gain the competitive advantage in today's fiercely competitive global market, companies must launch new products with higher quality in a shorter time. Meanwhile, the workload of assembly tasks is getting heavier and heavier with the production cycle becoming shorter and shorter, which puts forward higher demand for the workers. Assembly process is the last cycle of product to the market, the assembly quality will directly affect the brand image of the company. Manual assembly is still the most viable method for assembly task. The quality of a product largely depends on the experience of worker. The efficiency and healthy of workers may be affected because of the unreasonable assembly scheme or working in the uncomfortable condition for a long time, which will likely lead to error operations, low efficiency. One of the most popular methods to effectively grapple with that problem in advance is to carry out virtual assembly simulation 1. However, the application of virtual assembly is basically limited to process control and physical simulation stage, research on the simulation of interactive virtual assembly operations has been few studied to date.

In recent years, virtual assembly technology has enjoyed an explosive growth. From the beginning to now, it has been a model of industrial design in the manufacturing industry and has gradually been applied in aerospace products, ships, automobile manufacturing and other fields 2. Therefore, the article is to discuss the application of DELMIA software in the simulation of assembly operations and the improvement of the operating environment and study the assembly process planning with digital people involved, where worker operation should be taken into consideration to provide a more practical reference for assembly process design. By the construction of an ergonomic evaluation method for simulation of virtual assembly operation, a comprehensive analysis can carried out including physical workload, psychological load, posture comfort, etc. to make the analysis and evaluation of virtual assembly easier for users, which plays great roles in application and practice in improving the overall design level of products, the man-machine adaptation capability of equipment and shortening the development cycle.

People raise more and more concern about ever the technology of assembly evaluation in quality which involves in Man, Machine, Material, Method and Environment. In terms of life cycle of product, must give priority to the simplification of structure by the assembly of the parts and its processing characteristics, so the assembly quality evaluation is focused on the assemblability evaluation of the product.

In terms of the analysis of virtual assembly ergonomics, George Chryssolouris et al. made a research on the relationship of the layout of operation and its intensity to ensure good operations on the basis of the lifting equation of NOISH (The National Institute for

Occupational Safety and Health) and the energy expended model of Gragg 3. With regards to the judgement of fatigue status of digital humans through operational force analysis, Allen et al. 4 quantitatively evaluated the operation force and posture of manual assembly by virtual assembly technology and at the same time, they also implemented the research of the maximum assembly force and the average assembly force for each assembly cycle in order to avoid the repeat labor. On the evaluation of working postures, Mohamad et al. 5 completed the task by CATIA software to evaluate the postures of packaging industry workers, and they found that heavy objects and bending in operations have bad effect on the comfort of operation, which could be checked and improved by RULA gesture assessment.

2 Research on operation analysis of virtual assembly

This section gives priority to analyse the characteristics of the virtual assembly in environment and common human issues and discusses the application of DELMIA software in the simulation of virtual assembly and the improvement of the operating environment. First of all, in order to simulate the assembly, the entire assembly simulation can be completed through the building of virtual assembly environment by DELMIA software to build the virtual operation. Then, a simulation study is performed on the typical operations to discover and improve the irregular operation in the process in advance.

2.1 Simulation platform

The digital lean manufacturing application DELMIA (Digital Enterprise Lean Manufacturing Interactive Application) introduced by Dassault Systemes in France is characterized by strong simulation. As a subsystem for simulation of manufacturing and maintenance (PLM), DELMIA is able to perform simulation analysis of the entire manufacturing and maintenance in important equipment fields such as aerospace and defense, shipbuilding and automotive. The platform of the digital process planning, DPE has the capability of establishing product data, resource data and process structure to evaluate process plans and output results of assembly process. DPM is able to verify the simulation through the process flow of DPE and assembly process simulation. The digital human model is aimed to analyze the product design and ergonomic project. Ergonomic avatars are created as soon as possible with the DELMIA/Human module.

2.2 The decomposition for virtual assembly

Typical assembly operations are simulated with the DELMIA software to build a virtual assembly task. Based on the actual assembly task and motion analysis theory, the assembly operation hierarchy structure diagram is established, and the assembly operation is decomposed into assembly operation activity layer to form a series of basic assembly action units to realize the action simulation of the assembly task.

Operation Analysis (OA) is focused on analysis including objectives and methods of operations, work environment that affect the its efficiency and quality. According to different research objects, operation analysis is classified into four categories, as shown in Figure 1. The simulation in the report refers to that of the operation process in accordance with the assembly task, including the hands handling material, hand tightening fasteners and the lifting operation of components by the equipment 6.As for the virtual assembly, it should be studied by virtual ergonomic workers, virtual assembly tools, products to be assembled and lifting mechanisms to analyze the relationship among the operators, operations and work environment to improve the efficiency of assembly operations.

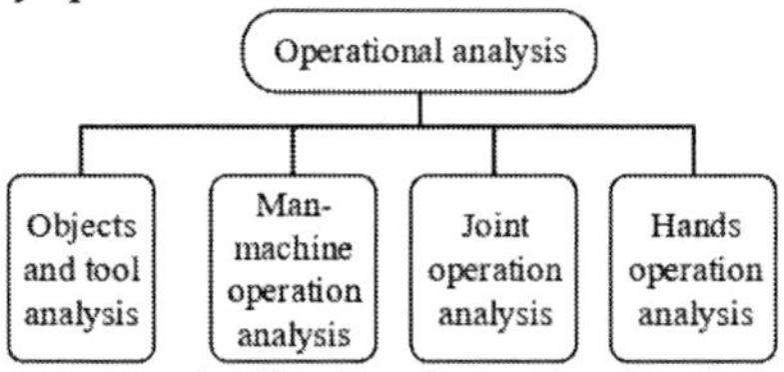

Figure 1. Classification of Operational Analysis

The hands operation analysis is mainly to avoid hand operation to improve work efficiency. Man-machine operation analysis is designed to make full use of energy balance of tools and personnel.

2.3 Motion analysis

Motion Analysis (MA) mainly studies body movements in the course of operation to eliminates unnecessary movements so as to formulate an optimal operation sequence and standard operation method 7. The American scholar Gilbuss discovered that the artificial operation consists of 18 motive factors and put forward the method of factor analysis that decomposes operation actions into basic action elements. Based on the theory of concentric circles proposed by Professor Zhou Dao and the impact of various factors on operations, these motive factors can be divided into inactive motives unrelated to operations, supporting therblig and effective therblig in the promotion of work 8.

1) Effective motives in the promotion of work is made up of common therblig and core ones with 8 categories. As seen in the Table 1.

Table 1. List of common therblig and core therblig

Category	Name	Instructions
Common therblig	Stretch out	Leaving or closing the objectives
	Hold	Holding objectives in arms or figures
	Move	Changing the direction of objectives
	Loosen	Loosing thins
	Preposition	Preposition before being fixed
Core therblig	Assembly	Combining two objectives
	Use	Changing objectives by tools or device
	Disassemble	Disassembling combined objects

2) Supporting therblig: it is obtained by effective therblig without direct relationship with operation. As an auxiliary factor, it should be cancelled or shortened as much as possible. The motion consists of 6 elements, as shown in Table 2.

Table 2. List of supporting therblig

Category	Name	Instructions
Supporting therblig	Examine	Comparing the product to standard one
	Look for	Searching the location of the objective
	Choose	Choosing the movements
	Plan	Considering the next action
	Position	Being in right place
	Find	The movements founded by objective

3) Invalid therblig: It has nothing to do with the operation, a delaying element, and must try to cancel them, which is focused by us, as seen in Table 3.

Table 3. The list of invalid therblig

Category	Name	Instructions
Invalid therblig	Holding	Keeping objective static
	Stop	Stopping only
	Delay	Caused by objective factors
	Avoidable Delay	Caused by subjective factors

2.4 Analysis of actions in key processes

The therblig reflects the movement characteristics of the assembler and the assembly operation can be decomposed into basic therblig, which constitutes the most basic operation movement unit of the assembly. Because of different shapes, sizes and weights, the description of the process is also different, but according to the real function, it can be roughly divided into feeding, positioning, assembly, connection, marking, inspection and discharge. assembly simulation should be facilitated on the basis of different combinations of therblig. The basic assembly operation is mainly related to that of the product without the adjustment of the posture in the course of assembly. The simulation of digital assembly is in separate with the simulation of adjustment of the posture according to the needs of the operation. On the basis of the therblig classification method of ergonomics and the requirements of the virtual assembly simulation for therblig, this paper takes the posture adjustment and position adjustment into consideration as basic therblig elements, composing the basic therblig of virtual assembly. The classification and definition of therblig are shown in Table 4.

Table 4. Basic therblig of virtual assembly

Category	Name	Instructions
Basic therblig of virtual assembly	Move	Changing direction
	Grab	Catching the objective
	Carry	The change of direction of objective carried by people
	Position	The alignment of tools and point, line and surface of the objective
	Place	Place objective in the location of assembly
	Examine	Examining whether the installation is appropriate
	Pose	Adjusting until the posture is appropriate
	Assembly	Installing subjective in right way
	Hold	Without change in posture and direction
	Tool	Installing by tools
	Release	Releasing tools or subjective after operation

At present, the analysis of the assembly process is only focused on assembly operations without therblig that better shows the characteristics of operators. For example, in the procedure of robot assembly, the fastening of the guide rails by screw, the assembly operation unit includes walking to the work station, grabbing the fastening tool, sending it to the assembly hole, aligning the top tight, tightening the screws, changing the working posture and placing wrenches 9. This paper proposes to decompose the assembly, which will reflect the characteristics of the operator in movement. The specific operation is shown in Figure 2.

The assembly event refers to one or more operations, such as electrical wiring and inspection and commissioning due to a fault or scheduled plan. The assembly operation is a part of it, including the installation of components of the robot, the disassemble and installation of the connector and the like 10. The basic assembly belongs to a work unit of decomposition of the assembly. For example, the assembly movement system can be decomposed into guide rail installation, tightening screws, mounting gaskets, the installation of mechanical stopper and limit switches and horizontal plates installation. The therblig of assembly means specific action of the assembler to complete the assembly. For instance, the fastening of the guide rails by screw, the assembly operation unit includes walking to the work station, grabbing the fastening tool, sending it to the assembly hole, aligning the top tight, tightening the screws, changing the working posture and placing wrenches 11.

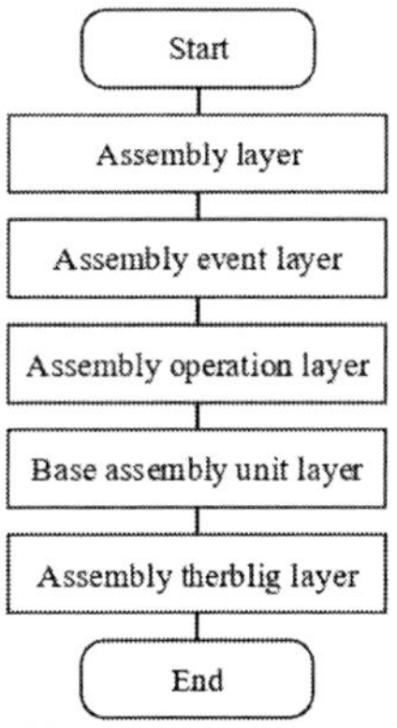

Figure 2. The hierarchy chart of assembly operation

3 Study on work efficiency evaluation method for virtual assembly

The section puts forwards to a comprehensive evaluation model based on multi-fuzzy comprehensive evaluation and analysis method of hierarchy, which provides a solution to the problem of work comfort evaluation for virtual assembly. The relevant theoretical methods and analysis of modeling process are described below in detail.

3.1 Ergonomics analysis and evaluation index system of assembly

Based on the simulation of the assembly operation, the comprehensive evaluation of the ergonomic analysis of the operator is designed to establish a corresponding index system. The selection of indicators is the basis for its establishment. Therefore, the selection of indicators is required for further research. First, a framework for evaluation indicators built by analyzing the ergonomic design standards and similar index system of ergonomic evaluation of assembly operations. In terms of the qualitative and quantitative requirements for assembly operations and the characteristics of the work environment of simulation assembly, preliminary evaluation indicators were selected and an index system suitable for ergonomic analysis of virtual assembly operations were established 12.

According to the characteristics of virtual assembly operations and the general steps of comprehensive evaluation, the process of establishing a comprehensive evaluation method of load-comfortability based on virtual assembly is summarized that consists of three parts. It's evaluation index, evaluation implementation methods and evaluation criteria. Based on the requirements for assembly, the selection of evaluation factors should be based on the feasibility of the integrated evaluation method applied in the simulation environment so that they are able to complete the analysis by instruments of simulation platform of the virtual assembly. The evaluation criteria are the core of the framework; the evaluation criteria in each evaluation index will have great effect on the accuracy of the evaluation. The main reference for the evaluation criteria refers to the description of requirements of the assembly operation based on the selected indicators 13. The specific chart of the evaluation of the comfort of virtual assembly operations is shown in Figure 3.

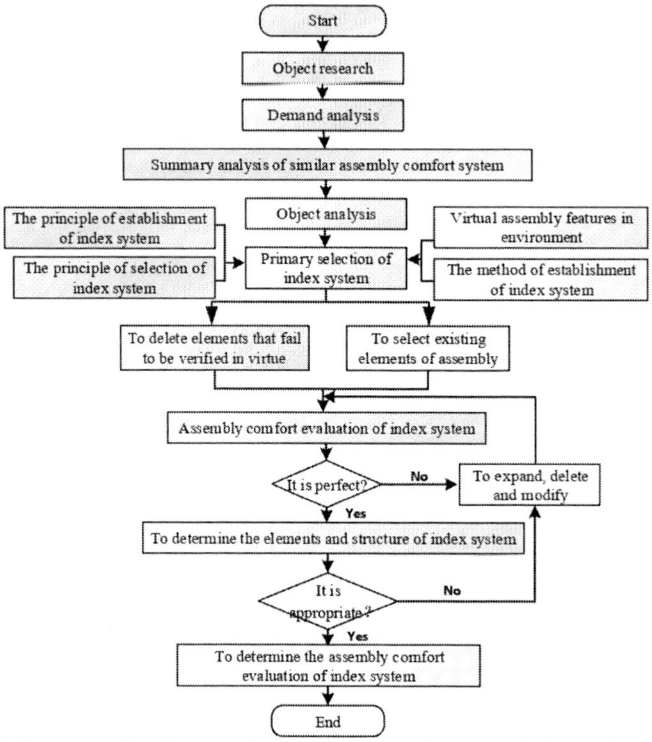

Figure 3. Flowchart for the construction of a comprehensive evaluation index system for the comfort of assembly operations

As shown in Figure 3, the analysis achieved by the characteristics of work environment of the virtual assembly and actual needs. Taking the existing research on the ergonomic evaluation index of assembly operations as a reference, a framework of evaluation indicators is initially constructed; with focus on the relevant requirements of the assembly operation, the evaluation index system initially screened out through the combination of the previous requirements analysis and the characteristics of the virtual assembly environment. Based on the preliminary indicator system, the index system can be expanded, deleted or modified by analyzing and judging whether the evaluation index system is perfect and whether the index can be verified under virtual conditions; the previous basic evaluation index system is classified, analyzed and gathered to establish the elements of assembly comfort evaluation index, among which each one is made up of a number of basic indicators. Each element and its supporting indicators are part of ergonomic evaluation index system of the assembly. It is necessary point out that since the evaluation index elements are inseparable from basic indicators, the latter is re-extended, deleted and modified when the evaluation index elements are incomplete, unreasonable in its structure or insufficient in its support. Then, the framework is re-established by a new basic index system.

The influences on operation comfort include improper posture, work intensity and duration, environment, psychological load, degree of match between tools and equipment and people 15. This article defines the comfort of the load: in the course of the assembly operation, the operator can succeed in completing the task within the specified time without discomfort, where he bears load including psychological, physical strength, posture and the environment. On the basis of the characteristics of virtual assembly operations in environment, the following evaluation indicators are selected: work posture, strength, noise, work content, work time and labor mode (carrying, picking, pushing and pulling).

Comprehensive analysis is carried out in the ergonomic requirements of assembly operations in a virtual assembly and a comprehensive evaluation index system for work comfort of virtual assembly to describe that the operator can succeed in completing the task within the specified time without discomfort, where he bears load including psychological, physical strength, posture and noise.

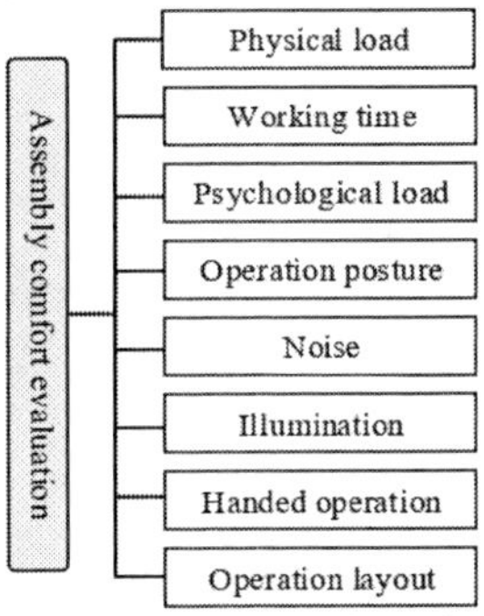

Figure 4. Comprehensive evaluation index system of comfort of load of assembly

3.2 Fuzzy comprehensive evaluation

Comprehensive evaluation refers to the evaluation of the complex system with multiple indicators which will integrate the indicator system, establish a comprehensive evaluation model and makes use of the model to evaluate and sort the objects 16. The choice of evaluation methods has a great influence on the its correctness. Different problems will have different evaluation method. When it comes to multi-objective evaluation, the evaluator should give priority to understand and analyze the evaluation object, the expected effect and the conditions. Then evaluation methods are screened out to ensure its effectiveness and objectivity according to the characteristics of the problem and the requirements for the evaluation.

Fuzzy Comprehension Evaluation (FCE) is based on the theory of fuzzy mathematics membership, in which it quantitatively transforms fuzzy indicators by construction of hierarchical fuzzy subsets and implements comprehensive evaluation on membership of multiple indices 17.The fuzzy comprehensive evaluation consists of six basic components: domain U, a kind of judging factor, set V related to the judging grade, matrix R, a fuzzy relation, matrix W, a kind of judging factor, the composition operator and vector S relevant to the evaluation result. Its basic principle: first, to determine the domain of the set $U = (u_1, u_2, \dots, u_n)$ and the set of domain of comment level $V = (v_1, v_2, \dots, v_m)$ for each indicator, among which u_i refers to individual evaluation index and v_j to its level of judgement. Then the weight of W and membership vector R for each indicator is determined. After fuzzy transformation, the matrix R of the fuzzy relation can be obtained. Finally, the matrix R and weight vector W are calculated by the fuzzy synthesis operator and normalized to get the result set S, as a result of comprehensive evaluation model made up of (U, V, R, W).

The membership function woks as the basis of fuzzy comprehensive evaluation. Different evaluation methods have different range of each evaluation index value and the value of the evaluation index can be mapped to the interval $[0,1]$ through the membership function to gain a quantitative and intuitive result of the evaluation index. The membership function form must be selected according to the requirements of evaluation goals. The two common forms of membership functions are shown in Figure 5.

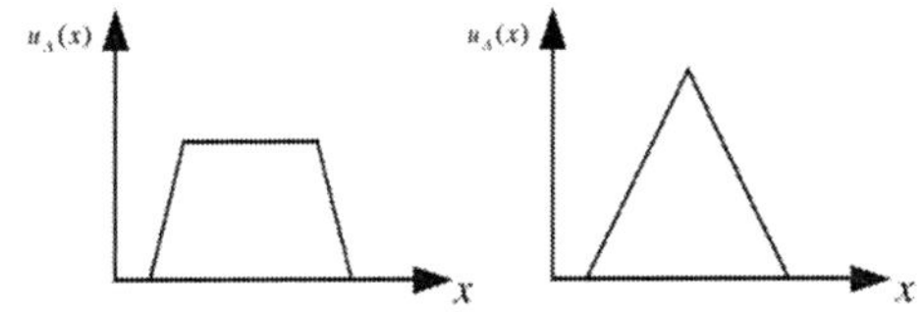

a)Trapezoidal form b)Triangle form

Figure 5. Diagram of the membership function

The core issue in comprehensive evaluation is to determine the weights in right way. When the evaluation system is more complex, it is difficult to directly give the weight of each evaluation index because of too many evaluation indicators and affecting each other; even though you do, the weight of each factor is very small to comply with the normalization, which is not conducive to analysis. Analytical Hierarchical Method (AHP) takes the advantage in determination of the weight of each evaluation index, so multi-level fuzzy comprehensive evaluation can matter 18. The specific application of Analytical Hierarchical Method is seen in the Figure 6.

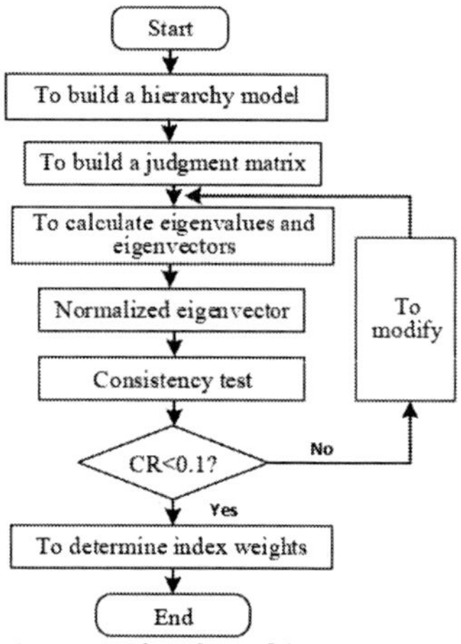

Figure 6. Flowchart of the AHP

1) Structure judgment matrix

First of all, it is necessary to analyze the questions and construct a hierarchical structure model. Then, according to the 9-level scale provided by F.L. Saaty scholars, the judgement matrix A is established by judging the importance of each level of evaluation indicators on the evaluation objectives. To compare the effect of n factors $X = \{x_1, \ldots, x_n\}$on the factor Z, two factor x_i and x_j was taken and a_{ij} is used to represent the relative importance of the influence of x_i and x_j on Z and a judgment matrix $A = (a_{ij})_{n\times n}$ is utilized to represent the comparison result, as shown in formula (1) and A serves as the judgment matrix between Z and X.

$$A = \begin{bmatrix} 1 & a_{12} & \ldots & a_{1m} \\ a_{21} & 1 & \ldots & a_{2m} \\ \ldots & \ldots & \ldots & \ldots \\ a_{n1} & a_{n2} & \ldots & 1 \end{bmatrix} \quad (1)$$

Among them, the value of a_{ij} is larger, if $a_{ij} > 0$, $a_{ii} = 1$, which shows the greater the importance of x_i to the evaluation target compared with x_j. The the impact of the importance ratio of x_i and x_j on Z is $a_{ji} = 1/a_{ij}$. 9 scales and their meanings are shown in the Table 5.

2) Calculation of eigenvalues and eigenvectors

The steps for hierarchical single sort: priority is given to get the max eigenvalue and eigenvector W of the judgment matrix A and normalize the W to obtain the sorting weight.

Table 5. Assignment Criteria of Elements in Judgment Matrix

a_{ji}	Definition
1	x_i and x_j are equal in importance
2	Between being equal and slight
3	x_i is slightly important than x_j
4	Between being slight and much
5	x_i is much important than x_j
6	Between being much and significant
7	x_i is significantly important than x_j
8	Between being significant and the most
9	x_i is the best important

3) Consistency examine of judgment matrix

In order to rank the importance of elements in the same problem, decision makers often suffer from the complexity of the problem and the subjective cause when looking at the problem, which results in discrepancies in the judgment results. To end this, it is necessary to adopt a consistent method of examination to determine whether the difference is allowed. The examination can be performed as follows:

a) Calculation of consistency index CI

$$CI = \frac{\lambda_{max} - n}{n-1} \quad (2)$$

b) In order to measure the consistency of the matrix, a Random Consistency Index (RI) is a must. The RI values of the 1st to 9th judgment matrix are shown in Table 6.

Table 6. The random consistency index value RI of the judgment matrix

n	1	2	3	4	5
RI	0	0	0.58	0.90	1.12
n	6	7	8	9	
RI	1.12	1.24	1.32	1.41	

c) Calculation of consistency ratio

$$CR = \frac{CI}{RI} \quad (3)$$

If CR<0.10, it is considered that the importance sorting of the judgment matrix is in line with the requirements. If CR0$\geq$10, the judgment matrix needs to be modified until the consistency check of the total sorting meets the requirements.

4) Total sort of final evaluation index system

Integrating the weights in the single criteria can get the total sorting weights. Let A and B be the upper and lower levels respectively, where $A = \{A_1, \ldots, A_m\}$, $B = \{B_1, \ldots, B_n\}$ and $a_1, \ldots, a_m$, represents the total weight of each element in the A-layer. The single sort weights of the B-level elements in the criteria are $b_{1j}, \ldots, b_{nj}$. Then, the total sorting weight of each factor of layer B, $b_1, \ldots, b_n$ can be obtained by formula (4).

$$b_i = \sum_{j=1}^{m} b_{ij} a_j, i = 1, \ldots, n \quad (4)$$

The accumulation of non-uniformity of the judgment matrix of each level may lead to inconsistency of the final total sorting weights, so the examination is implemented in total sorting of levels. The examination of the total sort random consistency can be calculated by equation (5):

$$CR = \frac{\sum_{j=1}^{m} CI(j) a_j}{\sum_{j=1}^{m} RI(j) a_j} \quad (5)$$

Among: CI(j),(j=1,…,m) refers to the consistency index with the corresponding judgment matrix a_j in the hierarchy;

$RI(j)$refers to average random consistency index with the corresponding judgment matrix a_j in the hierarchy;

If CR < 0.10, the level of total sorting ordering consistency is effective. If not, the weight CI values are obtained separately for the positive and negative matrices of different experts and the expert questionnaires whose CI values do not meet the requirements or the experts whose CI values is less 0.1 value are required to re-do questionnaires.

5) Croup decision-making

In order to overcome the subjective problem in determination of the weight of the analytic hierarchy process (AHP), the group decision-making of experts is a must and then aggregated to consider. The specific steps are shown in Figure 7:

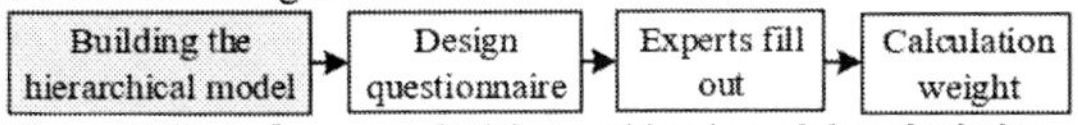

Figure 7. Steps for group decision-making in weight calculation

3.3 Fuzzy comprehensive evaluation procedure

Through the analysis method of AHP-FCE, the evaluation index fails to be obtained directly while gaining it by analyzing the problem and then constructing a hierarchical model. Firstly, the problem is decomposed based on the AHP, then the middle layer is used as the evaluation index which is the alternative. The specific steps for completing a multilevel fuzzy comprehensive evaluation by the AHP-FCE method are as follows:

1) To figure out the evaluation target U that refers to $U = (u_1, u_2, \ldots, u_n)$ in which n means the number of indicators.
2) To build a multi-level evaluation index system. Firstly, the evaluation objectives are decomposed based on the problem analysis to form the alternatives and criteria, as a result of a multi-level evaluation model;
3) To determine the weight W of the evaluation. As for the fuzzy comprehensive evaluation, the weights of the evaluation are calculated by the AHP: $W = (w_1, w_2, \ldots, w_n)$;
4) To determine the set V of the level of the evaluation: the evaluation target is divided in level according to the different evaluation indicators; for example, the evaluation of the comfort of assembly operations are established in level: $V = (v_1, v_2, \ldots, v_m)$;
5) To build a fuzzy relation matrix R. After construction of hierarchical fuzzy subsets is completed, a single-factor fuzzy relation matrix for each sub-indicator is gained:

$$R = \begin{bmatrix} r_{11} & r_{12} & \cdots & r_{1m} \\ r_{21} & r_{12} & \cdots & r_{2m} \\ \cdots & \cdots & \cdots & \cdots \\ r_{n1} & r_{n2} & \cdots & r_{nm} \end{bmatrix} \quad (6)$$

Among them, r_{ij} refers to the membership of the evaluation target u_i to the level fuzzy subset v_j.

6) To synthetize matrix S of result of fuzzy comprehensive evaluation. Although the principle of maximum membership is used commonly but not applicable to all. The wrong use will lead to great deviations in the evaluation results and fail to reflect the truthfulness of the evaluation of the problems. Therefore, it is possible to use a method that can sort the membership level by weighted mean according to their sort for a lot of things. fuzzy operations on W and R are implemented by a synthesis algorithm, vector S, the final result of fuzzy comprehensive evaluation is obtained:

$$S = W * R = (w_1, w_2, \ldots, w_n) * \begin{bmatrix} r_{11} & r_{12} & \cdots & r_{1m} \\ r_{21} & r_{12} & \cdots & r_{2m} \\ \cdots & \cdots & \cdots & \cdots \\ r_{n1} & r_{n2} & \cdots & r_{nm} \end{bmatrix} = (s_1, \wedge, s_n) \quad (7)$$

Among them, s_i refers to the membership of evaluated things as a whole to the fuzzy subset of v_j grade.

The section is to figure out the overall goal of the ergonomic evaluation for simulation process of the virtual assembly and compare and analyze the existing evaluation methods in a systematical method as well as determine the AHP-FCE combination to model the comfort of assembly operations, in which the AHP is used to determine the weights of indicators. Then further study is made in the method of fuzzy comprehensive evaluation and the fuzzy comprehensive evaluation system of assembly operation comfort is achieved, which provides a solution to the ergonomic evaluation of virtual assembly.

4 Conclusion

This paper studied the virtual assembly operation and took the workers into consideration in the assembly planning. It's provide a more practical reference for virtual assembly design of product. A systematic analysis of the issues related to virtual assembly and ergonomics was conducted. A set of research ideas from virtual simulation to ergonomics evaluation was proposed. An evaluation index system for evaluation of ergonomic analysis methods for virtual assembly operations was constructed, and an assembly for virtual assembly was proposed. The work efficiency fuzzy comprehensive evaluation method, in the fuzzy comprehensive evaluation model, the weight is determined using a strong logic, practicality and systematic AHP. The establishment of the model is in line with the actual situation, which facilitates the promotion of simulation analysis before product design or operation tasks, early detection of whether the assembly process scheme meets the results of human factors and gives suggestions for improvement, and improves existing designs to promote humans-machine-environment integration. The comprehensive evaluation model has the advantages of simple calculation, good practical application value and application prospect, so it is able to provide theoretical support for academic research and applications in industry.

References

1. Chryssolouris G, Papakostas N, Mavrikios D. A perspective on manufacturing strategy: Produce more with less[J]. CIRP Journal of Manufacturing Science and Technology, 2008, 1(1):45-52.
2. Jin Yongming. Research on Maintainability Analysis and Evaluation Method of Ship: Shanghai Jiao Tong University, 2011.
3. Chryssolouris G, Mavrikios D, Fragos D, et al. A virtual reality-based experimentation environment for the verification of human -related factors in assembly processes[J]. Robotics and Computer-Integrated Manufacturing, 2000, 16(4): 267-276.

4. Allen C, Karam K Z, Le Cam P, et al. Application of virtual reality devices to the quantitative assessment of manual assembly forces in a factory environment[C] Industrial Electronics, Control, and Instrumentation, 1995. Proceedings of the 1995 IEEE IECON 21st International Conference on. IEEE, 1995, 2: 1048-1053.
5. Mohamad D, Deros B M, Ismail A R, et al. RULA Analysis of Work-Related Disorder among Packaging Industry Worker Using Digital Human Modeling (DHM)[C] Advanced Engineering Forum. 2013, 10: 9-15.
6. Ming C L, Elmaraghy H A, Nee A Y C, et al. CAD model based virtual assembly simulation, planning and training[J]. CIRP Annals -Manufacturing Technology, 2013, 62(2):799–822.
7. Yao Y X, Xia P J, Liu J S, et al. A pragmatic system to support interactive assembly planning and training in an immersive virtual environment (I - VAPTS)[J]. The International Journal of Advanced Manufacturing Technology, 2006, 30(9-10): 959-967
8. Han Hu. Research on Human-Computer Interaction Scene Control Technology for Virtual Assembly of Aircraft[A]. Proceedings of 2017 International Conference on Advances in Materials, Machinery, Electrical Engineering (AMMEE 2017) [C]. 2017:4.
9. ZHINAN QIN. Research on Immersive Virtual Assembly Based on Virtual Reality Technology[A]. Proceedings of The 2nd Asia-Pacific Management and Engineering Conference (APME 2016) [C]. 2016:6.
10. Alzuheri A, Luong L, Xing K. Ergonomics design measures in manual assembly work[C]// Second International Conference on Engineering Systems Management and ITS Applications. IEEE, 2010:1-6.
11. Botti L, Mora C, Regattieri A. Integrating ergonomics and lean manufacturing principles in a hybrid assembly line[J]. Computers & Industrial Engineering, 2017, 14 (3) :350-355.
12. Rossi d, Bertoloni E, Fenaroli M, et al. A multi criteria ergonomic and performance methodology for evaluating alternatives in "manuable" material handling [J]. International Journal of Industrial Ergonomics, 2013, 43(4): 314-327.
13. Yang Q, Wu D L, Zhu H M, et al. Assembly operation process planning by mapping a virtual assembly simulation to real operation[J]. Computers in Industry, 2013, 64(7):869-879.
14. Zhao J. Study on the Application of the Virtual Technology in the Mechanical Assembly Process[C]// International Conference on Mechanical Engineering and Intelligent Systems. 2015.
15. Usman S, Zhu H T, Haq M U. Design Improvement of Assembly Workplace through Ergonomic Simulation and Analysis Using DELMIA[J]. International Journal of Engineering Research in Africa, 2016, 21:238-246.
16. Botti L, Mora C, Regattieri A. Application of a mathematical model for ergonomics in lean manufacturing[J]. Data Brief, 2017, 14(C):360-365.
17. Zhao Shuping, Liu Hongyang, Sun Gang Et al. Spacecraft assembly safety evaluation model based on fuzzy comprehensive evaluation [J]. Spacecraft Environment Engineering, 2014.,23:218-226
18. Ye Zhen. Study and Application of Fuzzy Comprehensive Evaluation Basedon AHP [D]. Guangzhou: South China University of Technology Guangzhou, 2010.

Aspects regarding the Conception, the Practical Realization and the Use of the "ETC - Mines" Tracked Mini Robot Functional Model in Humanitarian Demining Operations

Petrişor Silviu Mihai

"Nicolae Bălcescu" Land Forces Academy, Technical Sciences Department, 3-5 Revoluţiei Street, 550170, Sibiu, Romania

Abstract. In the present paper, the author want to highlight the construction and the possible implementation of a functional tracked mini-robot model within special humanitarian demining operations, a technological product developed within the *Advanced Logistic Technologies Laboratory* of our institution. The objectives and the hypotheses of the research, the methods and the instruments used in the process of the practical realization of the mechanical structure of the tracked mini robot as well as certain aspects related to its implementation and operation in humanitarian demining operations, a stage that was based on a study regarding the possible explosion by sympathy of a category of anti-tank mines, are presented in detail. We also mention the fact that the practical realization of the functional model of the "*ETC - Mines*" mini robot represented the starting point for the filing at the State Office for Inventions and Trademarks, Bucharest, Romania, of the patent application no. *a 2017 00562* representing the organological construction at a 1:1 scale of an innovative robotic technological product necessary for humanitarian demining missions.

1 Introduction

The usefulness of robotic technologies in operations of humanitarian demining of fields can no longer be questioned due to the innumerable advantages that they offer to the engineers in the theaters of operations (flexibility, minimal energy consumption using renewable energy sources, modular construction, robustness, small size, audio and video equipment, etc.) directly protecting them from being exposed to the hazards associated to the operations in question or to the detection of improvised explosive devices (IED) or of explosive devices, Explosive Ordnance Disposal (EOD) missions. However, the robot structures used in humanitarian demining operations are not only used in times of conflict but also in peacekeeping missions, all of these being part of the *United Nations Mine Action Program* (UNMAS) within the Department of Peace Keeping Operations, in countries where there have been military confrontations such as the Democratic Republic of Congo, Lebanon, Libya, Somalia, Sudan, etc. In all these states, the implementation of robust and flexible advanced technologies capable of demining mine fields without the direct involvement of the human factor and, subsequently, the replacement of the classic manual demining or by using oversized technologies to produce intentional detonation of mines are being attempted. The humanitarian demining of minefields represents a process of neutralization or destruction of the mines that can be found in a certain area or objective, this thing leading to the clearing of the affected land and it's rendering to the administrative authorities responsible for putting it into use. The humanitarian mine clearance operation is considered to be successful and the land is considered safe to use if 99.6 % of this is achieved [1].

2 Study on the explosion by sympathy of the main categories of anti-tank mines

The present study highlights the virtual explosion by sympathy for the destruction of a buried anti-tank mine using the trinitrotoluene explosive (TNT), this representing the most commonly used material in engineering operations. In this respect, we used a 3D simulation program in which detonation products in Euler-Godunov environments were introduced, and the buried mine is considered to be constituted of Lagrange-type elements, this being covered with a thin layer of polycarbonate. The model thus conceived is considered complete and the barrier conditions on the outside of it are of the Flow Out type thus allowing for the phenomenon to take place under similar conditions and outside these. In the simulation hypothesis (Figure 1), the following characteristics imposed by the existing environment of the action: a mine buried at 0.5 m with 9 kg of TNT explosive, an external environment considered air and two sensors that measure its pressure were taken into consideration. In order to carry out the explosions by sympathy, we used the model of Lee

Tarver-TNTCASTJJ1 material and during the simulation we followed the initiation by sympathy of the ALPHA variable, which represents the proportion between the detonating and non-detonating products, its value being 1 if the detonation is complete.

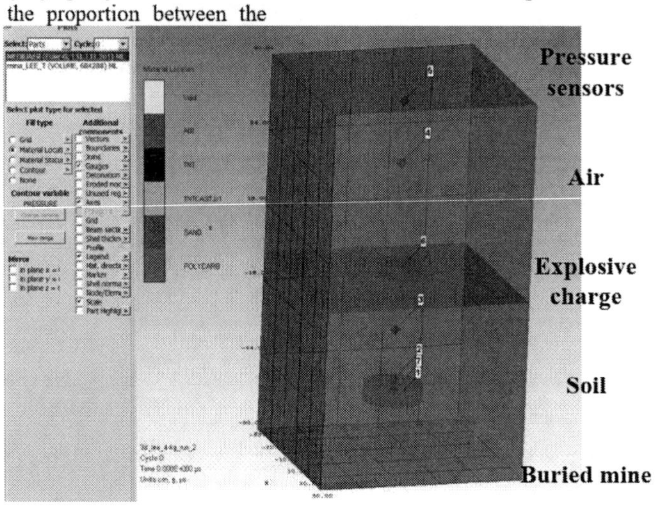

Figure 1. Presentation of the work environment.

Simulation no. 1: 4 Kg TNT. In the case of the first simulation, we used a quantity of 4 kg of TNT explosive in a cylindrical shape disposed over a 9 kg mine buried at 500 mm above the ground (Figure 2). At the moment 1.129 ms the mine begins to initiate and the so-called *hot spots* that demonstrate its initiation by sympathy appear (Figure 3).

Simulation no. 2: 1 Kg TNT. In the case of this simulation, a relatively small amount of TNT was used to test the behavior of the mine. At the moment 2.0 ms the mine is not initiated, thus one can consider that 1 kg of TNT disposed over the mine does not initiate it, the value of the ALPHA variable recording a small value (Figure 4).

Simulation no. 3: 1.17 Kg TNT. Since in the case of the previous simulation, the TNT quantity was insufficient to initiate the mine at 0.5 m from the surface of the earth, the amount of TNT was gradually increased until the mine was initiated, thus reaching the amount of 1.17 kg (Figure 5).

The conclusion of the simulations indicates that the quantity necessary to initiate a mine of about 9 kg buried at 500 mm in the earth should be between 1 and 4 kg. The exact quantity can be determined for each mine, depending on its characteristics, and a number of simulations similar to those briefly presented above should be made. On the basis of the above, steps could be taken for the practical construction of the tracked mini-robot model intended for the operation of humanitarian demining.

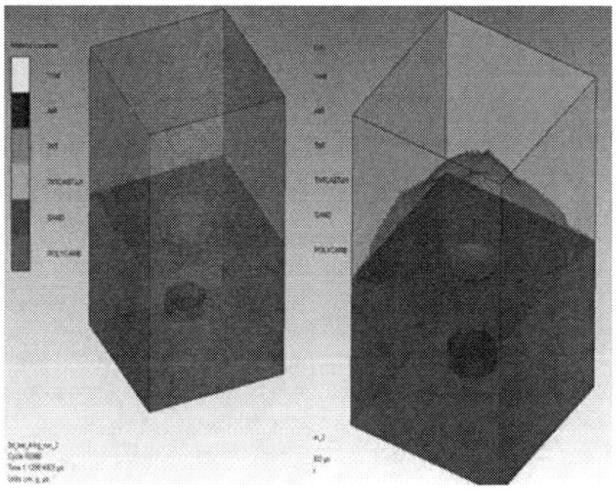

Figure 2. The explosion at 0.125 ms and at 1.129 ms respectively.

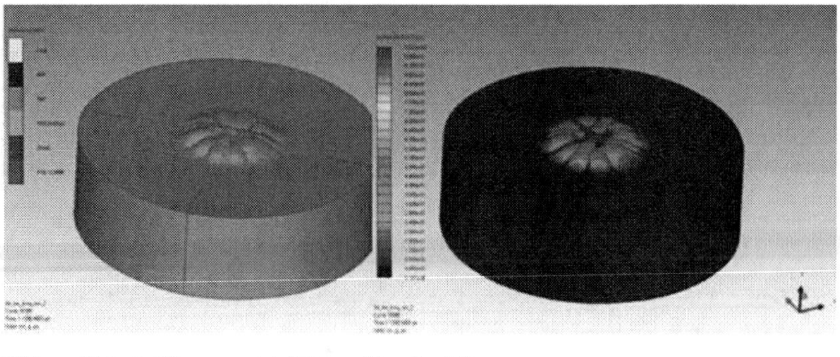

Figure 3. The ALPHA coefficient at the moment 1.129 ms of the simulation.

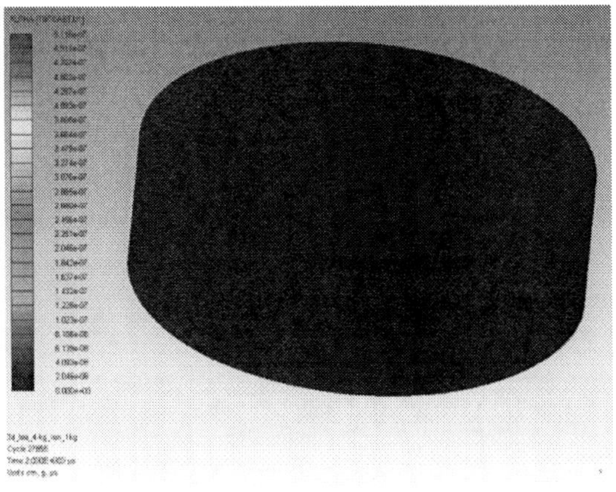

Figure 4. The ALPHA coefficient at the moment 2.0 ms of the simulation.

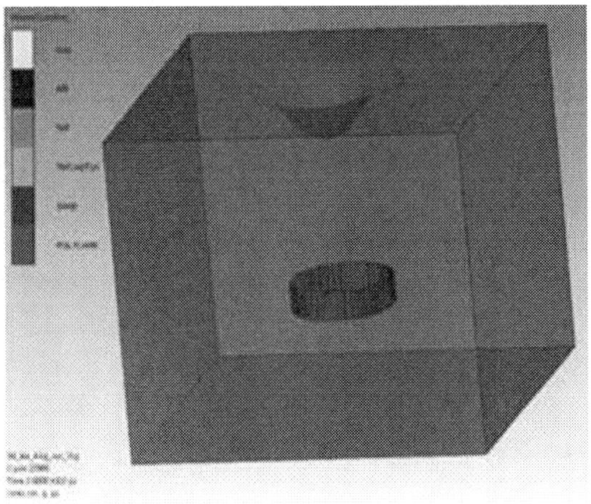

Figure 5. Explosion at 2.0 ms.

3 Concept and practical realization of the functional model of the *"ETC - Mines"* tracked mini-robot

In order to conceive and make the tracked mini-robot model intended for humanitarian demining operations in varied terrain (smooth or rough) during the time allocated for each mission, the following functional characteristics imposed to this were considered: to search and scan the land, to mark the lanes made in the minefields and to be capable to destroy the mines autonomously, to have a relatively low manufacturing cost (about USD 300) and a demining speed of at least 1.5 m/s, tobe able to funtion on the basis of solar energy being provided in this respect with photovoltaic cells that preserve energy in order to ensure uninterrupted operation, at the normal parameters of the minirobot organological configuration, both during daytime and at night. The mini robot locomotion is achieved by means of a smartphone and a bluetooth master slave module, the chassis of the structure is made up of two tracks with five carrier rollers driven by a driving wheel mounted by means of screws on its platform with a 40 cm opening (Figure 6). The driving wheel and the carrier rollers are fixed to an aluminum plate by means of a 2 cm long spindle. The tracks, 32 cm long and 5 cm wide, are made of plastic and are made up of 79 joints and oriffices necessary to move the mobile translation crew of the base of the mini robot (Figure 6). The *ETC-Mines* mini robot is powered by two 12 V DC motors that take energy from some photovoltaic cells and are controlled by a dual driver module, which also includes an Arduino Leonardo mother board (Figure 7) that helps to control the engine turation by means of the existing pins. Three ultrasonic distance sensors able to measure the distances up to a possible obstacle were attached on the mechanical structure of the mini robot model. By means of the programming code, the sensors will stop the mini robot at a set distance from the obstacle, and then they will help the mini robot by pass it. In order to achieve the placement of the trotyl packs on the ground, two arms that carry the packs of trotyl from the storage chamber of the mini robot by means of a mechanically driven conveyor belt were manufactured and attached to the mechanical structure of the mini robot. The conveyor belt is driven by a motor that mechanically transmits the motion to the drive drum towards the conveyor belt.

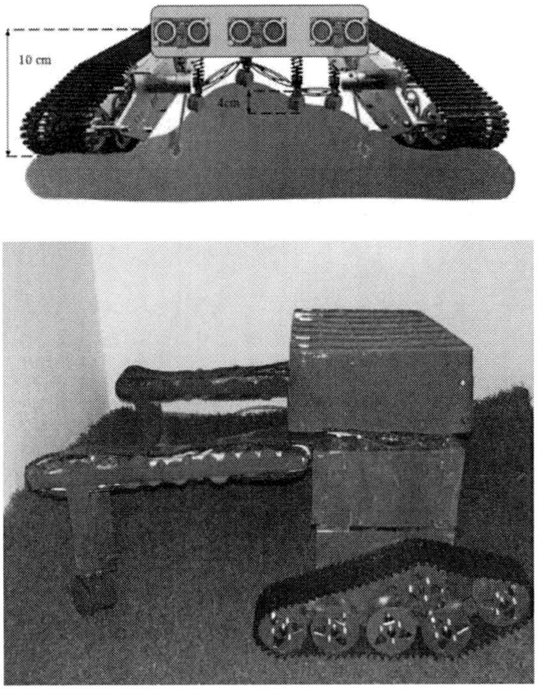

Figure 6. ETC – Mines mini robot– functional model.

In order to program the mini-robot structure (of the autonomous module), a programming language specific to C++ was used and in order to highlight the cleaned lane, the functional mini-robot model will mark the terrain during demining. In this sense, the mini robot was also provided with two red light led's simulating the marking of the lanes (Figure 8 a)).

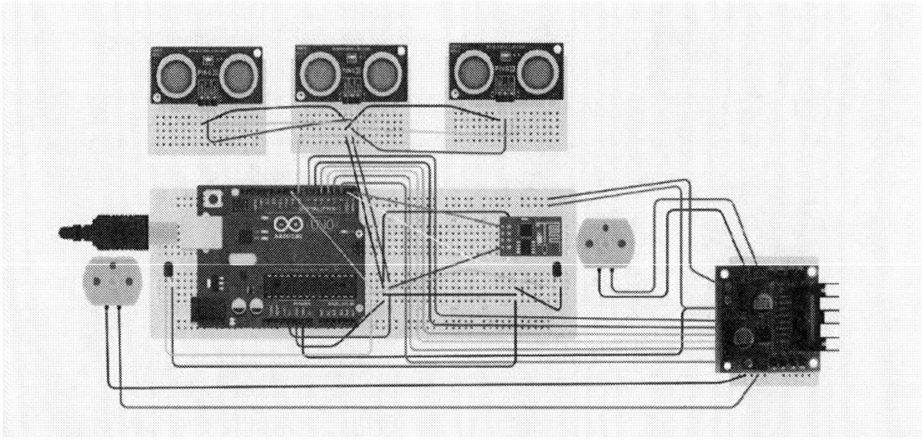

Figure 7. Scheme of the electronic components of the *ETC – Mines* mini robot.

This function is useful if it is vital to clear a lane in the minefield in high-risk areas such as CBRN - contaminated areas or areas that are under direct enemy fire. The function is not effective for achieving total mine clearance of a mine field, in this sense the wireless control performed by an engineer operator is needed. The mini robot can be remotely controlled by using an Android device that can make a bluetooth connection with a recognition device mounted on the mechanical structure of the model. The fact that the scanning of the mined ground can be done by means of a simple or infrared camera mounted on the mechanical structure of the mini robot is also mentioned; this camera transmitts, in real time, information to the engineer operator who is at the control desk and who, after the processing, can order the minirobot structure to carry out the demining of the field of anti personnel mine sand to avoid the obstacles encountered by the model. Figure 8 b) shows a scheme of total demining using wireless control.

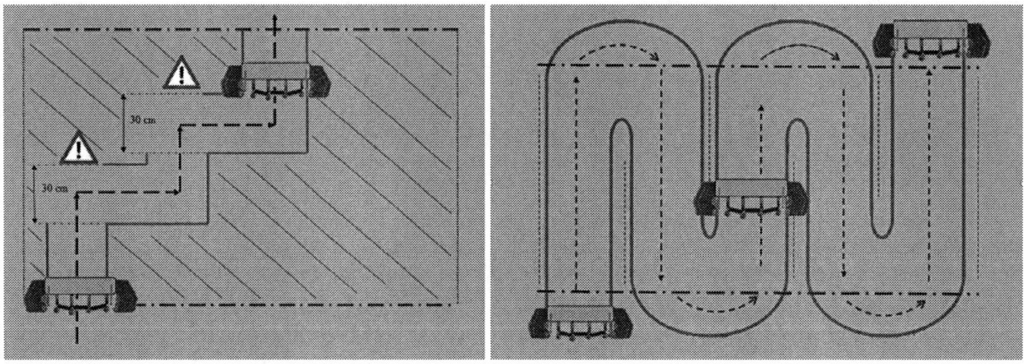

Figure 8. a) Scheme of the lanes in the mine fields using the "*obstacle-avoiding*" C++ programming; b) Scheme of total demining of mine fields using wireless control of an *ETC - Mines* mini robot operator.

4 Conclusion

Taking into consideration the advantages and the disadvantages of the robots used during the missions of demining mine fields, the author believe that their implementation within the demining groups that provide missions to ensure the mobility of the friendly troops is vital. This aspect is particularly highlighted by the need for safety and security of the demining groups in high risk areas. Thus, the preservation of the human resources is the most plausible argument that makes it possible today to rethink missions in accordance with the help received from robotic innovative products.

5 Acknowledgment

Participation with this paper was funded a grant type *PN-III-P1-1.1-MC-2017, No. 153* by *UEFISCDI*, Romania.

References

1. M. Habib, *Mine Clearance Techniques and Technologies for Effective Humanitarian Demining*, Journal of Mine Action, Issue 6.1, April, 2002.
2. R. Doare et all, *Robots on the battlefield – Contemporary issues and implications for the future*, Kansas, Combat Publisher Studies Institute Press US Army Combined Arms Center Fort, 2014.

The Lean Management of Spare Parts in Automotive Manufacturing

Guo Dongdong[1] ,Yu Xingwu[2]

[1]*No.8 Boxing Road, Economic Technological Development Area (BDA), Beijing, China*
[2]*No.8 Boxing Road, Economic Technological Development Area (BDA), Beijing, China*

Abstract. Spare part management is one of the most important work for enterprises, especially for manufacturing enterprises; however, the spare part management problems trouble enterprise operators a lot. In this article, implementation methods of lean spare parts management are illuminated. Spare parts purchase process is declared to reduce the purchasing cost and inventory value. We had established a suitable lean spare parts inventory management model for consumable parts, wear parts, insurance parts and accident parts. In addition, methods of lean spare parts management had been created base on optimized supply chain, ERP and integrating repeated material inventory. We used SAP-iPro system and self-developed system to manage spare parts, so that warehouse management process, spare parts purchase process and maintenance process are standardized. According to theory analysis and practice, the remarkable economic benefit is created for enterprise by the means of optimizing spare parts distribution, standardizing and scientific spare parts management.

1 Introduction

Nowadays, a higher level for the normal operation of production equipment has been required in the field of automotive manufacturing with the increasingly severe competition in the global market. There is no doubt that the downtime caused by equipment failure will bring huge losses to the enterprise. As a kind of indispensable supply for equipment maintenance, it is quite significant to keep enterprise's spare part inventory within a certain reasonable quantity [1]. Therefore, implementing scientific management for spare parts in the aspects of procurement, delivery date, warehouse maintenance, inventory control, allocation and so on will play an important role in the equipment maintenance work.

However, the surplus spare parts stock due to improper management will increase the operation cost of the enterprise. According to statistics, the spare parts inventory fund usually occupies about 7% of the total corporate liquidity. It has been identified that high cost of spare parts inventory has become the most troublesome problem in controlling the manufacturing production cost and production efficiency [2]. At present, in order to ensure the scientific purchasing and reduce the stock funds effectively, more and more enterprises have realized that it is quite important to implement the lean management of spare parts. There is no doubt that the lean management mode will be imperative in the field of spare parts management.

2 The overview of spare parts lean management

The development of lean management is based on Toyota's definition which called "just-in-time (JIT)" initially. This concept can be regarded as an operation mode to eliminate waste production continuously, which can satisfy the development goals of the enterprise [3]. Lean management mainly includes two aspects, "streamlined" and "beneficial". Lean management can not only reduce nonrenewable resources such as capital investment and resource, but also improve the economic benefit.

The lean spare parts management is a kind of enterprise management mode which is used to minimize the occupation of production resource and reduce the operating costs. On the one hand, this management mode highlights that it controls the spare parts consumption, ensures enough spare parts inventory, reduces the downtime due to lacking of spare parts; On the other hand, in order to reduce the inventory fund and liquidity as much as possible, this management mode also encourages to keep reasonable inventory quantity of the spare parts.

In recent decades, many international companies are starting to implement the lean management mode. At the same time, a large number of research achievements have been achieved in this field. For instance, Pedro José [4] has studied the relationship of the lean management, the supply chain management and the sustainability. Jared Ross [5] has discussed the successful lean management cases in the field of automobile manufacturing industry, the post office and health-care. The researcher also highlighted that it was quite significant to convene all the employees to involve in the project. Combining with the lean management theory, Xu [6] has studied the lean management approach which was suitable for the power supply enterprises. Wei [7] has researched the applications about the after-sale lean spare parts inventory management mode in Automobile Corporation. Xu [8] has not only evaluated the spare parts safety inventory and purchase plan but also developed the model of spare parts inventory management with the method of normal distribution.

3 Research status of lean spare parts management

Some certain classification work is needed to get the targeted features of every spare part. As is known to all, the main purpose of the spare parts management should find an appropriate balance among supply, working capital and operating costs. Different spare parts management strategies are suitable for different kinds of spare parts with different characteristics and supply chain processes. As a result, the spare parts classification methods have been widely used in the lean management so far.

The abroad lean spare parts management classification approach. The spare parts classification method is implemented earlier abroad. In the 1960 s, according to the properties of repairable, Allen and D'esop [9] had divided the spare parts into two categories roughly. Since then, in order to meet the actual needs, much more detailed spare parts classification methods have been raised. For example, R.Dekker [10] had divided the spare parts into critical and non-critical demand ones, and suggested to keep certain inventory for the critical parts. Considering the constraint of spare parts inventory, outage cost, safety problem, environmental goals, maintenance strategy and logistics, Braglia [11] had divided the spare parts into different categories and Cohen [12] had divided the spare parts according to urgent needs or just common demand.

The domestic lean spare parts management classification approach. The domestic researches on the spare parts classification are mainly using A/B/C classification methods. For instance, Jin [13] had developed a method to divide the spare parts by the A/B/C classification, combining with another classification according to their price, delivery time, quantity and lifetime. According to the cost proportion of each variety part, Liang and Wang [14] had divided the total parts into three categories (A/B/C) roughly, and then considered the use of fuzzy decision-making method to carry out a comprehensive evaluation of the detailed classification. Guo [15] had divided the spare parts into four categories (A/B/C/D) based on the degree of impact to the production, the regular pattern of failure and supply channels efficiency. Cui and Luo [16] had defined the repair availability, critical degree and economical index as the classification criteria, and classified the spare parts into A/B/C categories with the method of analytic hierarchy process.

4 The model of lean spare parts Inventory management

The mathematical model of spare parts inventory management can be established to keep the inventory scientific, so that spare parts can just meet with the maintenance demands without waste based on the lean model. According to actual production experience and spare part requirement, we have divided the total inventory into four categories. They are consumable parts, wear parts, insurance parts and accident parts as shown in Table 1 respectively.

Table 1. Spare parts classification methods

Classification	Characteristic	Target Number	Target value	Risk
Consumable parts	The probability of fault is 100% and the frequency of replacement is very high.(e.g. the frequency of replacement less than 1 year)	45%	15%	100%
Wear parts	The probability of fault is 100%, but the frequency of replacement is not so high. (e.g. the frequency of replacement more than 1 year)	35%	25%	100%
Insurance parts	The probability of fault is from 20% to 100%, and will be used after 5 years.	15%	30%	95%
Accident parts	Normally, these parts will never be used except the reason of quality, design, accident and so on.	5%	30%	unknown

The lean inventory management model of consumable and wear parts. Assuming that the lead time is a constant for each spare part, we have established the safety stock formula for consumable parts and wear parts which are shown in formula (4). These formulas [17] provide an important reference to the structural optimization for the inventory work.

$$\bar{d} = D / 365 \quad (1)$$

$$R = \bar{d} \times L + S \quad (2)$$

$$Q = \bar{d} \times T \ (T \geq L) \quad (3)$$

$$S = z\sqrt{\sigma_d^2 \times \overline{L} + \sigma_L^2 \times \overline{d}^2} \approx z \times \sigma_d \times \sqrt{\overline{L}} \quad (4)$$

Among the formulas above, D represents the annual demand quantity, R represents the reorder point, $\overline{d}$ represents daily average demand quantity, L and $\overline{L}$ are on behalf of the lead time and the average lead time respectively, S represents the safety stock, Q represents the quantity of replenishment, T represents order interval, Z represents the number of standard deviations under a certain service level, σ_d and σ_L are respectively represent the standard deviation of daily average demand quantity $\overline{d}$ and the lead time L . According to the inventory management model, we can get the scientific stock data, see in the table 2.

Table 2. Example of spare part inventory model

Material Number	Description	$\overline{d}$	L	S	R	R0
MCNP999920005687	Sucker	0.1232877	120	0	15	20
F303381075730	Silencer	0.0191781	120	0	3	4
MCNP999940003448	Pressure switch	0.0821918	50	0	5	6
QEV111AHJWCR	Pressure switch	0.0191781	120	0	3	5

According to the warehouse inventory model above, the spare part purchase model shows in figure 1. Figure 1 shows the relationship between inventory and consumption. When the inventory falls to the reorder point R, it creates purchase order on a predetermined quantity Q. The inventory level continues to decline during the lead time, and reaches safety stock. At the same time, the purchase spare parts come and the inventory level will rise. This model mainly relies on the control of reorder point R and order quantity Q to meet the best demand of inventory and the lowest total cost.

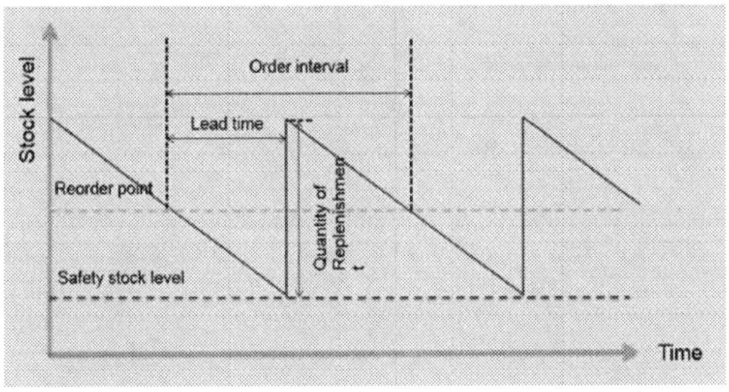

Figure 1. Relationship between inventory and consumption.

The lean inventory management model of insurance parts and accident parts. The Calculation model of the reorder point, the quantity of replenishment and the safety stock for the insurance parts and accident parts are shown in formula (5) to formula (7).

$$R = 1/2 \quad (5)$$

$$Q = 1/2 \quad (6)$$

$$S = 0/1 \quad (7)$$

where, R represents the quantity of replenishment, Q represents the quantity of replenishment, S represents the safety stock.

5 Methods of lean spare parts management

The lean spare parts management based on supply chain. Under the policy of the company to implement the lean management, we have established a new lean spare parts management model shown in Figure 2, which is based on the supply chain, such as controlling the procurement process, improving the supplier's quality management methods, implementing the centralized procurement and some other specific measures combined with the experience of the other large scale manufacturing enterprises spare parts management.

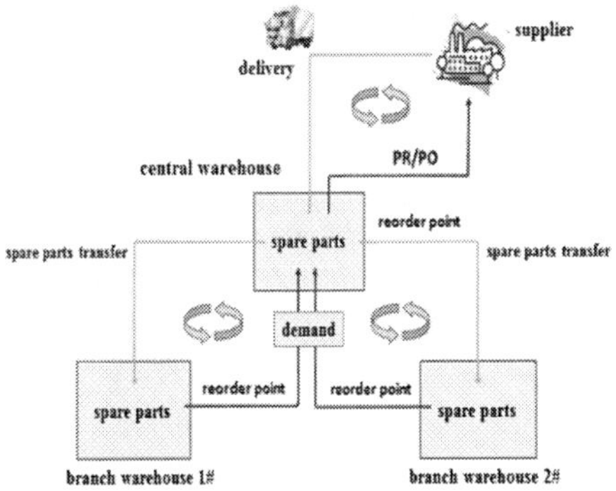

Figure 2. Schematic diagram of lean spare parts management mode based on supply chain.

The central warehouse has put into use after the hardware procurement, personnel training, process specifications and other preliminary preparation. By defining the logistics center as the function orientation, it has improved the cooperation of three maintenance team (factory maintenance workshop, MRA I repair workshop, engine plant maintenance workshop) significantly and make contribution to the standardization management mode which consists of centralized procurement, concentrated delivery and warehousing, distribution and centralized management. In addition, the measures have played a positive role in saving spare parts procurement costs and reduce spare parts inventory funds.

Central warehouse takes the advantages of centralized management for spare parts. Large amounts of spare parts stocks of brand-warehouses have been transfer to central warehouse, which optimizes warehouse stock quantity and stock value. The process of brand-warehouses' spare parts transferring to central warehouse shows in figure 3.

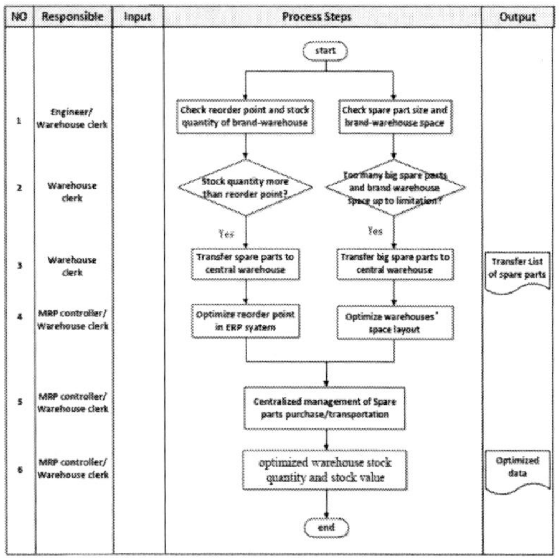

Figure 3. Process of brand-warehouses' spare parts transferring to central warehouse.

The lean spare parts management based on ERP. By building SAP-iPro intelligent ERP information management platform, We have completed the standardization of spare parts procurement processes, spare parts maintenance process and spare parts warehouse management process, the department also has organized the other departments of the enterprise and realized the sharing of information based on the data resource management system which plays a key role in the field of spare parts maintenance and management at the same time. The shortcomings of using Excel spreadsheets cause low efficiency, high error rate and sharing problems of the information, which has been solved by using the ERP information platform. There is no doubt that the implement of ERP system not only makes the company adapt to the demand of the modern enterprise better but also enhances the company's competitiveness with the modern management mode.

Up to now, the total cargo value of the ten branch warehouses and one central warehouse is about 2.25 billion RMB and involves about 33603 kinds of different spare parts. The SAP- iPro system and SAP-cbFC system have successfully been used in the management. In addition, combined with the spare parts maintenance demand and management requirements, the engineers of central maintenance department have developed spare parts management system software independently. And it has been realized to run on the tablet PCs, mobile phones and so on. This software realizes either the real-time spare parts order status or the key spare parts inventory information by the integration of information and data, the real-time spare parts inventory information can be reflected accurately, which facilitates the work of the related departments greatly. Figure 4 shows SAP-iPro and independently developed system for spare part management.

Figure 4. SAP-iPro and independently developed system for spare part management.

The lean spare parts management based on integrating repeated materials. In an optimal situation, different materials should have different type number, order number and the supplier number; the same kind of material must use the unified material information. But as a result of the practical work of variability and material information management is not normative, which generates a lot of repetition of material. The repeated material means the same kind of materials with different material numbers, each material information may be the same or different.

According to the feedback of MRP Controller and warehouse management personnel, material attribute identification is mainly composed of type number and order number, the possibility of repeat materials are summarized as follows.

1) Type number and order number are the same, but the material number is not the same, such as the following table 3

2) Type number and order number are filled in system without standard, maybe they are inclusion relation, such as "6005-2RSH" and "SKF:6005-2RSH".

3) Symbols of type number and order number are not uniformed, such as "6005-2RSH" and "6005-2RSH".

There may be "-", "/", "_" and " ", which are mixed but their main contents are the same.

Table 3. Example of repeated materials

Material Number	Type Number	Order Number	Description	Manufacture	Supplier
MCNP9920002889	6005-2RSH	6005-2RSH	Bearing	SKF	AOEHANG
MCNP9980003928	6005-2RSH	6005-2RSH	GROOVEDBALL BEARING	SKF	GUEDEL

According to the characteristics of the above repeated material, an edit distance similarity algorithm is used for accurately and quickly finding all the duplicate data in the system.

Edit distance [18]: the edit distance between two strings r and s defines as ed(r, s) or ED(r, s)), in which three kinds of editing operation (deleting, inserting or substituting a single character string) are used to make string r change into string s. The similarity between string r and s depends on how many steps of the editing operation. The fewer steps, the more similar.

For instance, string r = "abcd" and string s = "abd", ed (r , s)=1.

Standardized edit distance: the standard edit distance of string r and string s defines as sed as follows:

$$sed(r,s) = 1 - \frac{ed(r,s)}{\max\{|r|,|s|\}} \tag{8}$$

The value of sed is [0,1], sed is closer to 1, indicating that the more similar between the two strings.

In order to solve the edit distance between strings r and s, the standard dynami c programming algorithm can be used to solve the problem. The time complexity of the algorithm is O (|r|*|s|), and the space complexity is O (min (|r|, |s|)).

We can use a matrix D with |r|+1 rows and |s|+1 columns to calculate two strings' edit distance, the initialization of D[i][0]=i, 0≤i≤|r|, D[0][j]=j, 0≤j≤|s|, then we can make iterative calculation of D[i][j], 1 ≤ i ≤ |r|, 0 ≤ j ≤ |s|, according to the following formula:

$$D[i][j] = min(D[i-1][j]+1, D[i][j-1]+1, D[i-1][j-1]+\delta) \tag{9}$$

Among them, if r[i]=s[j], δ=0; otherwise δ=1. The final D[|r|][|s|] is the edit distance between strings r and s.

For example, given two strings s= "ser good" and "ser goal", calculate the edit distance between strings r and s by the standard dynamic programming algorithm. Run the process as shown in the following figure, the final ed (r, s) =2, see in figure 5.

	Φ	s	e	r		g	o	o	d
Φ	0	1	2	3	4	5	6	7	8
s	1	0	1	2	3	4	5	6	7
e	2	1	0	1	2	3	4	5	6
r	3	2	1	0	1	2	3	4	5
	4	3	2	1	0	1	2	3	4
g	5	4	3	2	1	0	1	2	3
o	6	5	4	3	2	1	0	1	2
a	7	6	5	4	3	2	1	1	2
1	8	7	6	5	4	3	2	2	2

(a) Standard dynamic program algorithm

	Φ	s	e	r		g	o	o	d
Φ	0	1	2						
s	1	0	1	2					
e	2	1	0	1	2				
r		2	1	0	1	2			
			2	1	0	1	2		
g				2	1	0	1	2	
o					2	1	0	1	2
a						2	1	1	2
1							2	2	2

(b) Dynamic program algorithm based on threshold

Figure 5. Edit distance graph for dynamic programming.

By using the edit distance similarity algorithm to analyze the data of spare parts, it can find out all the duplicate materials and process them in the system, so as to further optimize the spare parts inventory value. The result shows that about 16 thousand duplicate materials were found. These duplicate materials are combined as unique material and warehouse inventory is deduced. After 10 months' working on processing duplicate materials, nearly 40 million RMB was saved. Figure 6 shows detail of duplicate materials processing result.

6 Summary

Lean spare parts management is a kind of effective operation to minimize the quantity of inventory and reduce operating costs for enterprises, which closely depend on the accurate information, the establishment of the spare parts management database and the other applications of project management tools. As an equipment maintenance and spare parts management department of modern automotive manufacturing enterprise, we have absorbed the lean idea into the enterprise management culture. Practice has proved that not only a new lean mode of spare parts management has been explored but the cost of spare parts purchasing and inventory has been controlled, which makes great contribution to the developing of scientific and standard spare parts management.

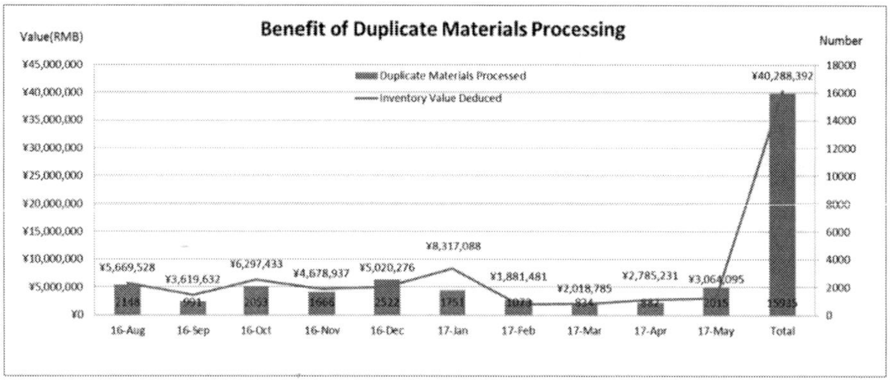

Figure 6. Benefit of duplicate materials processing.

References

1. DRIESSEN M, ARTS J, HOUTUM G J V. Maintenance spare parts planning and control:a framework for control and agenda for future research [J]. Production Planning & Control, 2014, 26 (5) , p. 407-426.
2. Zhang Haili. Research and Application of Spare Parts Inventory Optimization Mechod of A Steel Company [D].Zhe Jiang University,2013.
3. Lu Yongming. Research on the inventory management of Toyota Corporation service parts -- the application of lean management concept in service parts inventory management [D]. Nan Kai University,2011.
4. MART NEZJURADO, PEDROJOS Lean management: state-of-the-art and success factors in the adoption process. Evidence from the aerospace sector [J]. 2013.
5. FLYNN J R, VLOK P J. Lean Approaches in Asset Management Within the Mining Industry [J]. Lecture Notes in Control and Information Sciences, 2015, 20, p. 101-118.
6. Xu Yongteng. Lean management of Spare Parts in Power Supply Enterprise [J]. Guang DongScience and Technology, 2010, 19 (22), p. 82-84.
7. Wei Zhancheng. The Application of Lean Management in Chang He Automotive spare parts' customer service [D]. Nachang University, 2010.
8. Xu Weidong. Spare Parts Inventory Management Model based on Lean Model [J]. The Ten-day of Small and Medium-sized Enterprise Management and Technology, 2015, (3) , p. 36-37.
9. ALLEN S G, D'ESOPO D A. An Ordering Policy for Repairable Stock Items [J]. Operations Research, 1968, 16 (3), p. 669-674.
10. ARONIS K P, MAGOU I, DEKKER R, et al. Inventory control of spare parts using a Bayesian approach: A case study [J]. European Journal of Operational Research, 2004, 154 (3) , p. 730-739.
11. BRAGLIA M. Multi-attribute classification method for spare parts inventory management [J]. Journal of Quality in Maintenance Engineering, 2004, 10 (1) , p. 55-65.
12. COHEN M A, KLEINDORFER P R, LEE H L, et al. Multi-item service constrained (s , S) policies for spare parts logistics systems [J]. Naval Research Logistics, 2010, 39 (4) , p. 561-577.
13. Jin Xiwan, Ying Jingwen. Research on ABC Classification of Spare Parts Inventory [J]. Journal of East China Institute of Metallurgy, 1997, (1), p. 60-64.
14. Liang Guangtao, Wang Tianhong. Discussion on Classification Method of Spare Parts Inventory [J]. Construction Machinery Technology and Management, 1999, (5) , p. 28-29.
15. Guo Zhiming, Yan Hongsen, Chen Shihua, et al. Research on Spare Parts Inventory Control [J]. Computer Integrated Manufacturing Systems, 2003, 9 (6), p. 1028-1032.
16. Cui Nanfang, Luo Xue. ABC Classification Model Based on AHP for Maintenance Spare Parts [J]. Industrial Engineering and Management, 2004, 9 (6) , p. 33-36.
17. Wu Peili. The research of spare parts inventory management [D]. Harbin Institute of Technology, 2010, p. 14-18.
18. Yang Jian. Research on Top-k String Similarity Search Based on Edit Distance [D]. Harbin Institute of Technology, 2015, p.10-13.

AUTHOR INDEX